Table of Contents

The Stories

About the Stories

The 27 stories in *Read and Understand, Science, Grades 3–4* address science objectives drawn from the National Science Education Standards for grades K through 4. There are nonfiction and realistic fiction stories in the areas of life science, physical science, earth & space science, and science & technology.

When dealing with science content, certain specific vocabulary is necessary. This science vocabulary was discounted in determining readability levels for the stories in this book (which progress from low-third to high-fourth grade). A list of suggested science vocabulary, as well as other challenging words, is provided on pages 3 and 4.

How to Use the Stories

We suggest that you use the stories in this book for shared and guided reading experiences. The stories provide excellent opportunities to teach nonfiction reading skills, such as scanning for information and gleaning information from illustrations and captions.

Prior to reading each story, be sure to introduce the suggested vocabulary on pages 3 and 4.

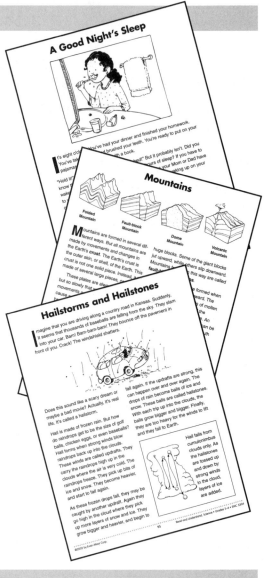

The Skills Pages

Each story is followed by three pages of activities covering specific skills:

- comprehension
- vocabulary
- a related science or language arts activity

Comprehension activities consist of two types:

- multiple choice
- write the answer

Depending on the ability levels of your students, the activity pages may be done as a group or as independent practice. It is always advantageous to share and discuss answers as a group so that students can learn from peer models.

Vocabulary to Teach

The content of the stories in *Read and Understand, Science, Grades 3–4* requires that specific science vocabulary be used. These words and additional words that your students may not know are given below. It is also advisable to read the story in advance to pinpoint any other vocabulary that should be introduced.

A Good Night's Sleep
pajamas, muscles, command, information, signals, mental, multiplication, scientists, confused, grumpy, clumsy, difficult

My Senses
senses, interact, environment, hearing, sight, touch, smell, taste, stomach, hungry, actually, terrible, engine exhaust, fountain, thirsty, organism, survive

Music to Your Ears
guitar, guitarist, vibrates, vibration, frequency, high-pitch, low-pitch, frets, tuning keys, melody, complex, chords, sound box, connected, imagination, sound waves, interprets

Not Just Dirt!
particles, layers, weathering, expand, contract, burrow, oxygen, earthworms, organic matter, temperature, fertile, organisms, bacteria, fungi, microbes, silt

Properties
properties, characteristics, objects, substance, physical, chemical, coarse, liquid, answering, describing, traits, observe, measure, compare, appearance

Playing to Learn
solitary, prey, stalking, cheetahs, retracts, pounce, special movements, antelope, behavior, practicing, connection, difference, wrestling, sharpening, survival

Make Your Backyard a Better Habitat for Birds
habitat, hummingbirds, nectar, identify, provide, foliage, several, completely, dissolved, refrigerator, prevents, fungus, appeal, unshelled, element, ceramic, attracted, accidentally, objects, probably

States of Matter
matter, element, atoms, molecules, combination, compound, liquid, sodium, chloride, hydrogen, material, definite, volume, pressure, inhaled, expand, water vapor, example, substance, observe, gradually, evaporates, oxygen

James B. Eads and His Famous Bridge
Mississippi, St. Louis, Civil War, engineers, successful, foremost, invented, remarkable, valuable, connected, surface, wealthy, Union, transportation, arches, inventor, creative, steel, vessel, cargo

Turned to Stone
sedimentary, limestone, skeletons, coral, dissolves, vinegar, quarries, igneous, metamorphic, formation, reacts, magnifying, Europe, ancient

Always Pointing North
compass, magnetic, molten, friction, core, cardinal points, intercardinal point, direction, Chinese navigators, compass card, lightweight, Pacific Ocean, Mount Everest, magnetic field

Fire in the Forest: Friend or Foe?
fertilizer, nutrients, habitat, naturally, litter, snags, fire fighters, community, dangerous, natural, lodgepole pine, exposed, non-native species, maintain, occur, prevent, advice, ecologists

The Story of Oil
petroleum, remains, material, separated, layers, refinery, fuel, gasoline, asphalt, plastics, medicines, rely, pollution, energy, alternative, renewable

Sally Fox: Spinning a Life

Rumpelstiltskin, biology, environment, resist, pesticides, organizations, achievement, natural, breed, fibers, dyes, inspects

Keeping Warm for Winter Fun

insulation, fibers, generates, prepares, chairlift, snowboarding, snowboarders, summit, plumes, furnace, fiberglass, chickadees, manage, lodge, slopes, flitting, Vermont, skiers, powder

Marc Hauser: Learning About Animal Minds

termites, humans, laboratory, broad, stage, screen, tamarin monkeys, rhesus monkeys, experiments, results, expected, sorrowful, especially

Planetary Almanac: Interesting Facts About Our Solar System

cycle, solar flares, spews, high-energy particles, satellites, space probe Ulysses, atmosphere, sulfuric acid, biomes, climate, conditions, temperate forest, exploration, ripples, Europa, astronomers, Mercury, Venus, Earth, Mars, Jupiter, Saturn, Uranus, Neptune, Pluto

When the Dragon Swallows the Sun

ancient, solar eclipse, astronomy, astronomer, celestial, Chinese, Babylonia, Babylonian, Egyptians, Ptolemy, total eclipse, partial eclipse, amazement, peaceful, dim, grave, dark, wonder, total, completely, tomb, terrified, viewing

Hailstorms and Hailstones

Kansas, hailstorm, updrafts, layers, hailstones, damage, destroyed, injure, thunderstorms, tornadoes, Wyoming, Colorado, Nebraska, Oklahoma, Texas, centimeters, flattened, shatter, crops, dent, slippery, pelting, hail, tropical

The Miracle of Light

energy, twinkling, lightning, zigzags, glimmering, aurora borealis, northern hemisphere, southern hemisphere, aurora australis, chemicals, fireflies, abdomens, squid, artificial, laser beams, entertainment, cables, Internet

A Class by Itself

species, features, classification, taxonomy, Aristotle, Carolus Linnaeus, system, mammals, rodents, cheetah, leopard, biologists, identify, traits

At Home in the City

wilderness, habitat, raccoons, adapt, coyote, garbage, predators, prey, rabies

Making Old Things New

recycled, recycling, landfills, garbage, decompose, environment, plastic, pollution, communities, process, recyclables, waste stream, recycling contractor, remolding, detergent

Mountains

crust, plates, erosion, folded mountains, faults, fault-block mountains, glaciers, crevices, dome mountains, South Dakota, volcanic mountains, molten, lava, vent, Hawaiian Islands, Black Hills

The Magic Eye

emergency, traditional, radiation, photographic, dense, computerized tomography, organs, blood vessels, Sir Godfrey Newbold Hounsfield, Allan Macleod Cormack, diagnose, tissues, tumors, medical, operation, patient, technology, fracture

Petrified Forest National Park

Arizona, swampy, volcanic eruptions, minerals, petrified, fossils, reptiles, dinosaurs, crystals, Jasper Forest Overlook, generations, preserve, park ranger

Nature's Gifts: The Materials of the Earth

weapons, materials, flexible, wander, natual materials, fiber, copper, iron, machinery, jewelry, alloy, chemicals, petroleum, inventors

A Good Night's Sleep

It's eight o'clock. You've had your dinner and finished your homework. You've taken a bath and brushed your teeth. You're ready to put on your pajamas and hop into bed with a book.

"Hold it!" you say. "It's too early to go to bed!" But it probably isn't. Did you know that most kids your age need about 10 hours of sleep? If you have to wake up early, you should go to bed early, too! Does your Mom or Dad have to wake you up each morning? If you often have trouble waking up on your own, you may not be getting enough sleep.

Why is sleep important? Sleep gives your body a chance to rest. Your muscles relax. Your heart slows down. Your body takes a break and builds energy for the next day. Sleep also helps your body heal when you are sick. It helps your body grow. And sleep is very important for your brain, too.

Your brain is the command center for your body. All day long your brain is hard at work. It takes in all kinds of information. It uses that information to tell your body how to respond. For example, on a hot day you might feel very warm. Your brain sends signals to your skin to make you sweat so that you'll cool off. As you're playing baseball, your eyes tell your brain that a ball is flying toward you. Your brain sends signals to your arms. You swing the bat and hit the ball! Your brain takes charge of mental tasks, too. It tells you how to spell words on your spelling test. It stores facts like your telephone number. It remembers the multiplication table. Every moment of every day, your brain is busy.

When you go to sleep, your brain does not "turn off." But at least it gets a break from taking in information. Some scientists think your brain does an important job while you sleep. They think it sorts the information it has taken in during the day. Some people believe that your brain can even solve problems while you sleep. That's why we say, "Why don't you sleep on it?" to someone who is worried or confused.

Your brain also spends some time dreaming every night. Scientists do not really understand dreams. They think that dreams might help your brain make sense of things that happen during the day. Dreams can be silly or scary, happy or sad. Often, they don't seem to make sense. In the morning, you may not remember your dreams at all.

What happens if you don't get enough sleep? For one thing, you'll be pretty grumpy. Missing sleep can make you clumsy, too. You are more likely to drop things or trip over your own feet. You won't be able to think as well, either. Even simple tasks are difficult for a tired brain. If you don't get enough sleep, you won't be at your best.

If you want to have a good day, start with a good night's sleep!

Name _____

Questions about
A Good Night's Sleep

Choose the best answer.

1. Sleep is important because _____.

   ○ it lets your body build energy for the next day

   ○ it lets your brain rest

   ○ it helps your body grow

   ○ all of the above

2. About how many hours of sleep are needed by most people your age?

   ○ 7

   ○ 14

   ○ 6

   ○ 10

3. How often do people have dreams?

   ○ about once a week

   ○ only when they are sick

   ○ every night

   ○ never

4. If you don't get enough sleep, you will most likely feel _____.

   ○ cheerful

   ○ grumpy

   ○ curious

   ○ full of energy

5. Which of these things does **not** happen when you sleep?

   ○ your muscles relax

   ○ your brain sorts information

   ○ your heartbeat slows

   ○ your lungs stop working

Vocabulary

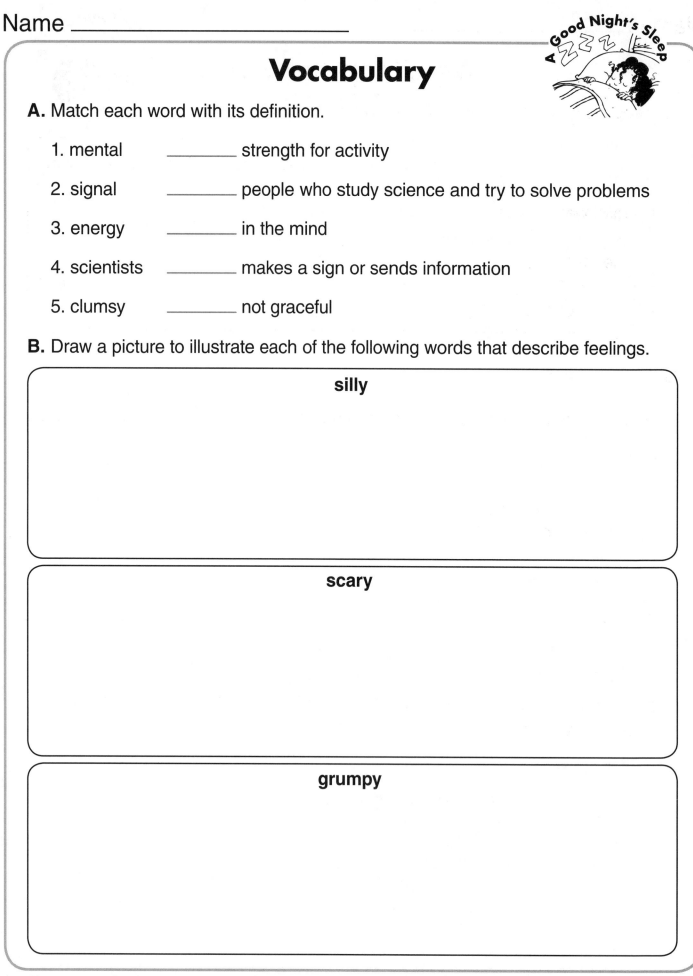

A. Match each word with its definition.

1. mental   _____ strength for activity

2. signal   _____ people who study science and try to solve problems

3. energy   _____ in the mind

4. scientists   _____ makes a sign or sends information

5. clumsy   _____ not graceful

B. Draw a picture to illustrate each of the following words that describe feelings.

silly

scary

grumpy

Name _____

My Sleep Record

A. Keep track of the amount of sleep you get for one whole week.

	Sun.	Mon.	Tues.	Wed.	Thur.	Fri.	Sat.
Time I went to sleep							
Time I woke up							
Number of hours I slept							

B. Answer the following questions about your sleep record.

1. On which night did you get the most sleep? _____

2. On which night did you get the least sleep? _____

3. Did the amount of sleep you got affect the way you felt the next day? Explain your answer.

   _____

   _____

   _____

4. Do you find that it is easy or difficult for you to get enough sleep? Why do you think this is so?

   _____

   _____

   _____

My Senses

"Aaron, time to get up," called his mother. "It's a school day."

Aaron sat up in bed. He turned on the light. He pushed the record button on the tape player. The machine made a low humming sound.

"Today I begin a science project for Mr. Martin's class," said Aaron. "Our assignment is to keep a 'sense' diary for a day. He wants us to record how we use our **senses** to **interact** with our **environment**. We are to describe how we use our senses to get what we need.

"The first sense I used today was hearing. I heard my mom calling first thing. It's funny, but I was already awake. I wake up at the same time every morning. I wonder how that happens? My body must get used to getting up at the same time every day.

"I used sight next because I couldn't see anything in the dark. I turned on the light. And the next sense I will use is taste. My stomach feels hungry. I need breakfast," said Aaron. He clicked off the tape machine. Then he picked it up and headed into the kitchen.

"Actually, I was wrong," said Aaron into the machine. "The next sense I used was my sense of smell. The toast smelled really good this morning. And it tasted good, too. I wonder if things smell and taste good when you're hungry?"

"Turn that off now please," said Aaron's mother.

Click.

"Now I'm riding on the bus to school," said Aaron. "There was a terrible engine exhaust smell just a minute ago. How can some things smell so bad and some smell so good?"

Click.

"I see Mr. Martin has moved all the desks," whispered Aaron into the tape machine. "He did the same thing a few weeks ago. Now my desk is closer to the window. He says he likes to change things in our classroom environment. The first time he did this we didn't know what to do. Now we know to look for the desk with our name.

"This got me thinking about my environment. I looked the word up in the dictionary. It says that an environment is a place where an **organism** lives and **survives**. My environment includes several places— home, school, and my grandma's house. I learn about different things in each place."

"Time to turn off the tape machine, Aaron," said Mr. Martin.

Click.

"It's almost time to go home now," said Aaron. "I've been too busy today to talk. But I wanted to talk about one more sense that I use all the time—the sense of touch. I pick up a pencil to write. I push the drinking fountain button when I'm thirsty. And I kick a soccer ball on the field for fun!

"I'm going to my grandma's house after school today. We are making tamales for the holidays. I'm sure I'll use all of my senses there. And I will continue using all of my senses to interact with all parts of my environment. I'm not sure I could stop it, even if I tried.

"The end."

Click.

My Senses

Questions about *My Senses*

1. How did Aaron know that breakfast was ready?

   _____

   _____

2. Tell at least four ways Aaron used his sense of touch.

   _____

   _____

   _____

3. Aaron used a tape recorder and the sense of hearing to tell his story. How might he have used the sense of sight to tell the same story?

   _____

   _____

4. Tell how you use each of your senses in your school environment.

   sight _____

   _____

   hearing _____

   _____

   touch _____

   _____

   smell _____

   _____

   taste _____

   _____

Name _____

Vocabulary

A. Use the words in the box to best complete the sentences below.

Word Box		
interact	environment	organism
sense	survive	

1. I used my _____ of sight to experience the lovely sunset.

2. A polar bear would not thrive in a tropical _____.

3. The scientist examined the _____ under the microscope.

4. I _____ with my friends at school.

5. Wild animals must know how to find food in order to _____.

B. Illustrate two of the sentences listed above. Write each sentence on the lines below the picture.

_____   _____

_____   _____

Name _____

Use Your Sense of Taste

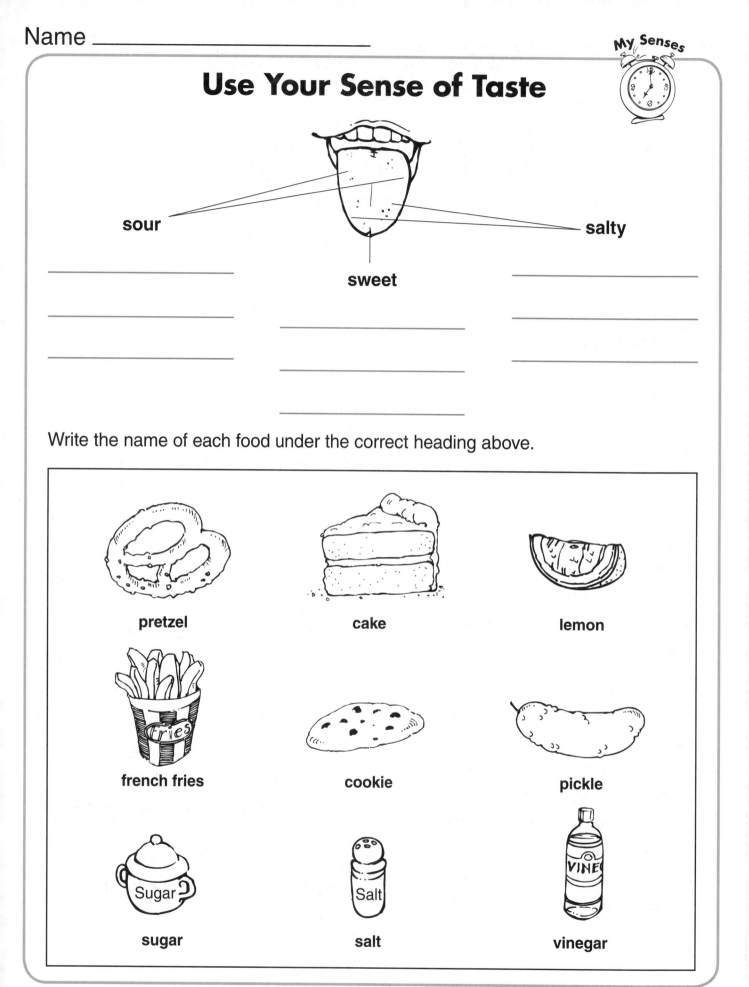

sour

sweet

salty

_____   _____   _____

_____   _____   _____

_____   _____   _____

_____

Write the name of each food under the correct heading above.

pretzel

cake

lemon

french fries

cookie

pickle

Sugar

sugar

Salt

salt

VINE

vinegar

Music to Your Ears

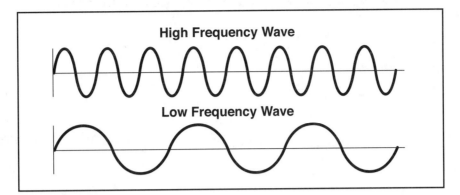

You have probably heard a guitar playing. Maybe you heard a guitarist at a concert. Or maybe you listened to a song on the radio. Do you know how the guitar created the music? The answer is waves of sound.

A sound is created when an object **vibrates**, or moves back and forth. These vibrations make **sound waves** that move the air around the object. When the sound waves enter our ears, our brain interprets, or understands, them as different sounds.

Every sound has a different **frequency**. Frequency is the number of sound waves that are created in one second. If there are a lot of waves, the sound has a high frequency. A high frequency creates a **high-pitch** sound. For example, a whistle has a high frequency. If the sound produces few waves per second, it has a low frequency and produces a **low-pitch** sound. A truck engine has a low frequency.

Now let's look at a guitar. A guitar usually has six strings. A person plays the guitar by plucking the strings. Plucking makes the strings vibrate. These vibrations create sounds.

If you look closely at a guitar's strings, you'll see that they are not all the same. Some are thin. Some are thick. The thick strings vibrate slowly. The thin strings vibrate quickly. This means that the thick strings make a lower-pitch sound than the thin strings.

A guitar's strings are connected to **tuning keys**. These keys let the guitarist change how tight the strings are. A tight string vibrates more quickly than a loose string. This means the tight string will make a higher-pitch sound.

But a guitar with six strings can make a lot more than six sounds! Making a string shorter can change the sound it produces. A guitarist does this by pressing his finger on the frets. Frets are special ridges on the neck of the guitar. Now the string will vibrate at a different frequency. That means it will make a different sound.

A guitarist can pluck each string by itself. This plucking creates a series of notes that can form a **melody**. A melody is simple, but it isn't always very interesting. To make more complex music, a guitarist plays **chords**. A chord is created when several notes are played at the same time. To play a chord, a guitarist places his fingers in different places on the neck of the guitar. Then he plays all the strings together. Each string makes a different sound. These sounds blend together to make music.

Have you ever held a rubber band between your fingers and plucked it? If you have, you know that the sound it makes is not very loud. The same is true of guitar strings. A vibrating guitar string does not make a loud sound. This is because the guitar string does

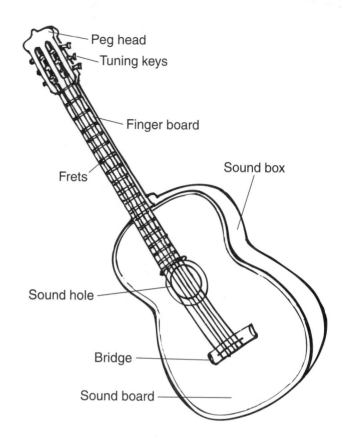

Peg head
Tuning keys
Finger board
Sound box
Frets
Sound hole
Bridge
Sound board

not vibrate hard enough to move a lot of air. Since only a small amount of air moves, the sound is hard to hear.

To solve this problem, guitar strings are connected to a wooden **sound box**. When the strings vibrate, the box vibrates, too. This creates a bigger vibration. That bigger vibration creates a louder sound.

Next time you are listening to a guitar, think about those sound waves moving through the air. If you use your imagination, you can almost feel the music playing!

Name _____

Questions about
Music to Your Ears

1. What creates a sound?

   _____

   _____

2. What is the frequency of a sound?

   _____

   _____

3. What kind of pitch does a sound with a low frequency create?

   _____

4. How does a person create sounds from a guitar?

   _____

   _____

5. How do tuning keys change the sound produced by individual strings?

   _____

   _____

6. What is the difference between a melody and a chord?

   _____

   _____

7. What does a sound box do?

   _____

   _____

Vocabulary

A. Read the dictionary entries for each vocabulary word below. Then choose the correct word to complete each sentence.

> **chord** *(KORD)* Noun. A combination of musical notes played at the same time.
>
> **frequency** *(FREE-kwuhn-see)* Noun. The number of sound waves created in one second.
>
> **fret** *(FRET)* Noun. A ridge on the neck of a guitar.
>
> **interpret** *(in-TUR-prit)* Verb. To understand or figure out.
>
> **melody** *(MEL-uh-dee)* Noun. A series of notes.
>
> **sound wave** *(SOUND WAVE)* Noun. A series of vibrations that can be heard.
>
> **vibrate** *(VYE-brate)* Verb. To move back and forth quickly.

1. Jean placed one finger on a _____ on the guitar.

2. The gong will _____ when you hit it.

3. I couldn't _____ his secret message.

4. Will played a _____ with three notes in it.

5. A _____ travels to your ear.

6. My mother hummed a pretty _____.

7. A whistle has a high _____.

B. Look up the word **guitar** in a dictionary. Then answer the questions.

1. What definition does your dictionary give for **guitar**?

_____

_____

2. What part of speech is the word **guitar**? _____

3. Does your dictionary have a picture of a guitar? _____

The Parts of a Guitar

Study the diagram of the guitar. Then fill in the blank in each sentence.

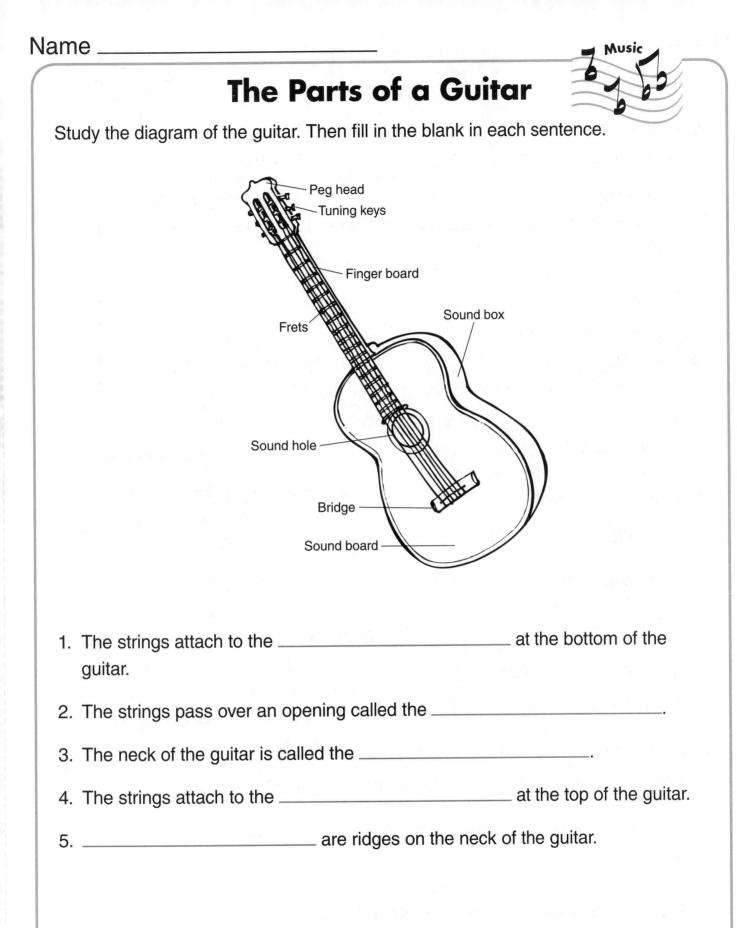

1. The strings attach to the _____ at the bottom of the guitar.

2. The strings pass over an opening called the _____.

3. The neck of the guitar is called the _____.

4. The strings attach to the _____ at the top of the guitar.

5. _____ are ridges on the neck of the guitar.

Not Just Dirt!

Earnest Earthworm here. I want to tell you about my favorite subject—soil. It's a whole lot more than dirt!

All soil starts as rock. Over many years, rocks are worn away. This process is called **weathering**. Many things cause weathering. One cause is **temperature** changes. When the temperature is hot, rocks **expand**. That means they get bigger. When the temperature is cold, rocks **contract**. That means they get smaller. Expanding and contracting breaks off pieces of the rock.

Sometimes water gets into cracks in the rock. When the water freezes, the cracks get bigger. Sometimes tree roots grow into cracks in rocks. Animals can make a crack bigger as they burrow, or dig, in the dirt around it. As the crack becomes bigger, more pieces of rock break off.

Water can also cause weathering when it flows over a rock. Flowing water slowly wears away layers of rock. All of these changes break rocks into smaller pieces called rock **particles**.

Rock particles mix with the **organic matter** (the remains of dead plants and animals), water, and air to form soil. Soil also contains some living **organisms** such as **bacteria**, **fungi**, and **microbes**.

- *It can take more than 500 years to form one inch of topsoil.*
- *There are over 70,000 kinds of soil in the United States.*
- *Up to 10 tons of animal life can live in an acre of soil.*
- *The spaces between soil particles are called pores—just like on your skin.*

Soils can be classified by the size of the rock particles they contain. **Clay** soil has the smallest rock particles. **Sand** has the largest rock particles. The size of rock particles in **silt** is larger than clay and smaller than sand. Soil with smaller particles can hold more water. Soil with larger rock particles can hold less water.

A soil's color can tell you a lot about it. In general, dark soil is more fertile, or richer, than light soil. **Fertile** soil can support more life than infertile soil. If soil is red, it has a lot of iron. Most plants grow well in this type of soil.

If soil is yellow, it may mean that water does not drain well from the soil, and the land is probably poorly drained. Gray soil is usually poor, too. The gray color shows that the soil has little iron or oxygen.

Soil may look like just a pile of dirt. But life on Earth, including yours truly, couldn't survive without it!

Earthworms Are Helpful

I may be small but I do a big job!

Earthworms spend almost all their lives digging through soil. As they travel through the soil, they move little pieces of it around. This creates new spaces in the soil. These spaces allow the soil to hold more water. They also allow more air to move through the soil. Soil that is full of earthworms is almost always healthy.

Name _____

Questions about *Not Just Dirt!*

1. Tell what part each of the following might play in weathering.

   a. temperature: _____

   _____

   b. water: _____

   _____

   c. animals: _____

   _____

2. What four things are found in soil?

   _____

   _____

3. How does the size of the rock particles in soil affect the amount of water it can hold?

   _____

4. Which of the kinds of soil mentioned can hold the most water?

   _____

5. How do earthworms make soil healthy?

   _____

   _____

6. If you wanted to plant a vegetable garden, which kind of soil would you choose?

   ○ yellow   ○ gray

   ○ light-colored   ○ dark-colored

   ○ red

Not Just Dirt!

Vocabulary

A. Write the bolded word from the story on the line beside its meaning.

1. small pieces

_____

2. soil that can hold a lot of water

_____

3. to get smaller

_____

4. wearing away

_____

5. rich, productive

_____

6. soil that can hold little water

_____

7. the "hotness" or "coldness" of something

_____

8. to get bigger

_____

9. to dig a hole in the ground

_____

B. Write a sentence using each word in the box.

Word Box		
earthworm	soil	rock

1. _____

_____

2. _____

_____

3. _____

_____

Not Just Dirt!

True or False?

Read each statement below. Then write a **T** next to the statements that are true. Write an **F** next to the statements that are false.

_____ 1. All soil starts out as rock.

_____ 2. Only changes in weather cause weathering.

_____ 3. Water is an important cause of weathering.

_____ 4. Soil includes rock particles, water, air, and nutrients.

_____ 5. Clay soil has large rock particles.

_____ 6. Sand can hold little water.

_____ 7. Plants can't grow in fertile soil.

_____ 8. Gray soil is the best soil for plants.

_____ 9. Dark soil is usually more fertile than light soil.

_____ 10. Earthworms are good for the soil.

_____ 11. An earthworm spends most of its life above the ground.

Properties

Every object on Earth
Can be simply described
By color or weight,
By shape or by size.

These characteristics
We call properties,
And they help us explain
Everything that we see.

We can ask a few questions
To help us define
The traits that belong
To each object we find.

Is it heavy or light?
Is it thick? Is it thin?
Is it hot, warm, or cold?
Can it wiggle and bend?

Is it liquid or gas?
Is it silky or coarse?
Is it as small as an ant
Or as big as a horse?

Is it round? Is it square?
Is it red, blue, or green?
Is it as hard as a rock
Or as soft as a dream?

When you are answering
Questions like these,
You are describing
Properties!

Properties Can Change

Did you know that the properties of an object can change? Some changes are **physical**. These changes make the object look different, but do not change the makeup of the substance the object is made of. For example, paper can be torn into shreds, but it is still paper. Metal can be bent or flattened, but it is still metal. Wood can be sliced into boards or sawed into sawdust, but it still is wood. A physical change affects the shape or appearance of an object. A physical change does not affect the makeup of an object.

But some changes are **chemical**. This means that the substances that make up an object are changed, and one or more new substances are formed. When wood is burned, it is completely changed. When a piece of wood burns in the fireplace, some of it turns into gases. These gases escape up the chimney. The rest of the wood turns into ashes. The makeup of the wood, as well as its appearance, has changed. This is a chemical change.

Notice all the things around you. What are the properties of the things you see? How can you use these properties to describe the different objects? Use your senses to explore the properties of things in your world.

Investigate Properties

All objects have certain **properties** that we can observe, measure, and compare. Work with a small group of classmates to do this investigation.

Gather these items:
- a coin
- a marble
- a bottle cap
- a key
- a leaf
- a rock
- a pencil
- a jar lid
- a paper clip
- an orange
- a rubber ball
- a postage stamp
- a sheet of paper
- a wooden block
- a safety pin
- an empty plastic bottle

Examine the properties of each of the objects you found. Sort the objects into groups by color. Sort again, this time for shape. Find all the objects that are made of metal. Which ones are made of wood?

Find two objects that are made of wood and compare them. Which one is larger? Use a ruler to find out. Which one is heavier? Measure their weights with a scale if you have one.

Questions about *Properties*

1. Properties can be _____.

   ○ observed

   ○ measured

   ○ compared

   ○ all of the above

2. A scale is a device that is used to measure _____.

   ○ length

   ○ temperature

   ○ weight

   ○ height

3. Which of the following is an example of a chemical change?

   ○ bending a paper clip

   ○ burning wood

   ○ tearing paper

   ○ sawing wood

4. A ruler or tape measure is a device that is used to measure _____.

   ○ weight

   ○ temperature

   ○ length

   ○ none of the above

5. When wood burns, some of the substance of the wood turns into _____.

   ○ ice

   ○ gas

   ○ paper

   ○ metal

Vocabulary

Characteristic means a quality, trait, or property of a particular thing.

A. Complete each sentence.

A characteristic of carrots is that they are _____.

A characteristic of kittens is that they are _____.

A characteristic of fire is that it is _____.

A characteristic of snow is that it is _____.

B. Draw a picture to illustrate each of the following characteristics.

bent	flattened
sliced	shredded

Name _____

Properties

Scavenger Hunt

List as many objects as you can that have both properties given in each example.

green and smooth	long and thin
short and thick	**round and yellow**
small and cold	**rough and bumpy**

©2002 by Evan-Moor Corp.   *Read and Understand, Science • Grades 3–4 • EMC 3304*

Playing to Learn

Have you ever watched kittens play? They chase their tails. They jump and pounce on each other. They grab a hanging string. They climb on each other and bite ears or paws.

All this play is a lot of fun. But it is also a way to learn. Playing teaches important skills. The kittens will need these skills later, when they hunt for their food.

A pet cat usually doesn't have trouble finding food. But cats that live in the wild have to hunt. Hunting is probably the most important skill kittens have to learn. One way they learn this skill is by playing.

Most cats are solitary hunters. That means they hunt by themselves. A cat catches its prey by stalking it. It creeps slowly toward its prey, hiding so the prey can't see it. When the cat gets close enough, it pounces. Then it bites its prey to kill it. Small cats eat mice, birds, lizards, and insects. Large cats eat bigger animals, such as wild pigs.

When kittens play, they show many things that are also done while hunting. One kitten might see its mother's tail waving back and forth. The kitten crouches down. Then it begins to creep forward. It is stalking Mom's tail! When the kitten gets close enough to the tail, it pounces. It lands on the tail and rolls over. Its paws are wrapped around the tail to hold it. It might even bite the tail. These movements—stalking, pouncing, grabbing, and biting—are very important. They are the same ones the kitten will use later when it is hunting.

Pet kittens also play with toys. Have you ever seen a kitten chase a ball or smack a rolled-up piece of paper? This behavior teaches skills that the kitten could use to catch a mouse or a cricket. Some kittens like to jump high to catch a feather on a string. If that kitten wanted to catch a bird, it would do the same jump.

©2002 by Evan-Moor Corp.

Some wild cats use special movements when they hunt. Cheetahs sometimes use one paw to hit their prey in the side and knock it down. Scientists have watched cheetah kittens play. They saw the playing kittens use this same paw-slap.

A lion cub often knocks another cub down by placing one paw on the other cub's shoulder. Adult lions sometimes catch antelope the same way. The lion places one paw on the antelope's shoulder. Then it pulls the antelope down. Once again, a play behavior teaches the lion a hunting skill.

Kittens aren't the only cats that play. Adult cats also play. You may have seen two adult cats wrestling. One cat will bite the other. Or it might smack the other cat with its paw. The cats look like they are fighting. But they are not. They are practicing their fighting skills and keeping their muscles in shape. Cats may also play together to make a connection with other members of their family or group.

Cats seem to know the difference between playing and fighting. When a cat plays, it retracts its claws. This means it pulls its claws into its paw. Retracting the claws keeps one cat from hurting the other. Cats also pretend to bite each other when they play. But their teeth don't break the skin. Just like people, playing is best when no one gets hurt!

It's Not Just Cats

Digital Stock

Cats aren't the only animals who like to play. Many wild animals also play. Some monkeys play hide-and-seek or king-of-the-mountain. Beavers wrestle and splash in the water. Seals play keep-away and tag with balls of seaweed. Bears chase and wrestle each other. A kind of bird called the kea hangs upside down from a branch and swings. All of these behaviors help animals by sharpening their survival skills.

Playing to Learn

Questions about
Playing to Learn

1. It is important for animals to play because _____.

   ○ it is fun

   ○ it helps them get along with each other

   ○ it keeps them from getting bored

   ○ it teaches important skills

2. One of the most important skills predator animals need to learn is _____.

   ○ fighting

   ○ playing

   ○ hunting

   ○ sleeping

3. Which food is **not** eaten by small cats?

   ○ birds

   ○ wild pigs

   ○ mice

   ○ insects

4. A paw-slap in the side is most often used by _____.

   ○ cheetahs

   ○ lions

   ○ tigers

   ○ pet cats

5. Another reason cats play is to _____.

   ○ have fun

   ○ keep in shape

   ○ get attention

   ○ find food

6. Bears practice fighting skills by _____.

   ○ swimming

   ○ biting

   ○ climbing

   ○ wrestling

Playing to Learn

Vocabulary

A. Write the number of each word on the line in front of its definition.

1. pounce   _____ to pull in

2. solitary   _____ hunting quietly and sneakily

3. prey   _____ making better

4. stalking   _____ alone

5. retract   _____ doing an act again and again

6. behavior   _____ the ability to stay alive

7. connection   _____ a coming together

8. sharpening   _____ an animal that is hunted by another animal

9. survival   _____ to jump

10. practicing   _____ a way of acting

B. Illustrate two of the words listed above. Label your drawings.

Playing to Learn

Learn About Cats

There are many different kinds of cats. Here are pictures and information about four different cats. Write the name of the correct cat on the blank.

| tiger | lion | cheetah | bobcat |

1. I live in Africa. Males, females, and cubs all live together in a large group called a pride. We hunt together, too. Males have long manes of hair, but females don't have a mane. I am a _____.

2. I am the fastest cat in the world. I can run about 60 miles an hour! I have very long legs and a long body. My fur is covered with dark spots. I am a _____.

3. I live in North America. My tail is very short. I live in many places including deserts, woods, and swamps. I am not as big as other wild cats. I hunt small animals such as rabbits, squirrels, and mice. I am a _____.

4. My stripes help me hide in the long grass when I am stalking prey. I usually live by myself. I am the largest wild cat, and I can be more than 9 feet (2.7 m) long. Most cats don't like water, but I like to swim! I am a _____.

Make Your Backyard a Better Habitat for Birds

An animal's habitat is the place where it lives. The habitat supplies all the things the animal needs to survive. These things are food, water, shelter, and a place to raise young.

Would you like to make your backyard into a better habitat for birds? You don't need a lot of space, or a lot of money. Start by looking over the outdoor space around your house or apartment. Make a list of the kinds of plants and trees you find there. Ask an adult to help identify them.

Do some of the plants or trees make berries that birds like to eat? Are there bright flowers that provide nectar for hummingbirds? Do any of the shrubs or trees have thick foliage that makes them good hiding and nesting places? Look at the chart on this page to find the names of plants and trees that birds use as homes or sources of food. Add some of these plants to your yard if possible. Some of the plants can even be grown in pots on a small deck or patio.

Provide more food for birds by setting out several kinds of bird feeders.

Hummingbirds feed on sweet nectar. You can make a sort of "homemade nectar" for them to eat. You will need to buy a hummingbird feeder. Wash and rinse it well. Ask an adult to help you make the nectar. Place 4 cups (950 ml) of water in a saucepan and bring it to a boil. Add 1 cup (200 g) of sugar and stir until sugar is completely dissolved. Let the mixture cool to

Tree or Plant	Food	Cover	Nesting
Cedar	X	X	X
Cherry	X		
Mulberry	X		
Holly	X	X	X
Pine	X	X	X
Crab Apple	X	X	X
Serviceberry	X		
Spruce	X	X	X
Elderberry	X	X	X
Honeysuckle	X		
Bee Balm	X		

©2002 by Evan-Moor Corp.

Read and Understand, Science • Grades 3–4 • EMC 3304

room temperature. Fill the clean feeder with the cool nectar. Store extra nectar in the refrigerator.

At least once a week, ask an adult to help you clean the feeder using water, a little bit of bleach, and a bottlebrush. This prevents the growth of fungus that can harm the hummingbirds.

Purchase or build a simple seed feeder. Black oil sunflower seeds appeal to many birds, so they are a good choice. Hang the feeder from a tree so that squirrels and other animals cannot steal the seeds. Clean these feeders with water and bleach as well. Do this every few weeks.

Here are some other interesting ways to feed the birds:

Tie a string to the top of a pinecone. Spread peanut butter all over the pinecone, pressing it into all the spaces. Hang from a tree branch.

Purchase a whole coconut. Ask an adult to help you crack it into large pieces. Tack a piece of the unshelled coconut to a board or tree branch.

Fasten apple or orange halves to the feeding station in the same way.

Water is another important element of habitat. You can easily make a birdbath. A large ceramic saucer that is used to catch water underneath a flowerpot makes a fine birdbath. Or use an upside-down garbage can lid balanced on rocks. Hang a plastic milk jug filled with water above your birdbath. Poke a tiny hole in the bottom of the milk jug. Birds will be attracted by the sound of the dripping water.

Birdbaths and bird feeders need to be up off the ground so that birds can more easily spot cats that might try to catch them.

Place feeders away from picture windows so that birds will not accidentally fly against the glass. Many birds are injured or killed in this way. Ribbons, stickers, and hanging objects help show the birds that they cannot fly through.

If you follow these steps, you will probably have many new bird visitors. Make sure to clean and fill the feeders often. Your feathered friends will thank you.

Name _____

Questions about
Make Your Backyard
a Better Habitat for Birds

1. What is a habitat?

   _____

   _____

2. What are the four important things that a good habitat must supply?

   _____

   _____

3. What do hummingbirds eat?

   _____

4. Why shouldn't birdbaths and bird feeders be placed on the ground?

   _____

   _____

5. Why is it important to keep hummingbird feeders clean?

   _____

   _____

6. What makes a shrub or tree a good hiding place for birds?

   _____

   _____

7. Name two plants that provide food, cover, and nesting places for some birds.

   _____

   _____

Name _____

Vocabulary

Backyard Habitat

Label the items in the picture using the words in the box.

Word Box		
berries	shrub	patio
birdbath	branch	foliage

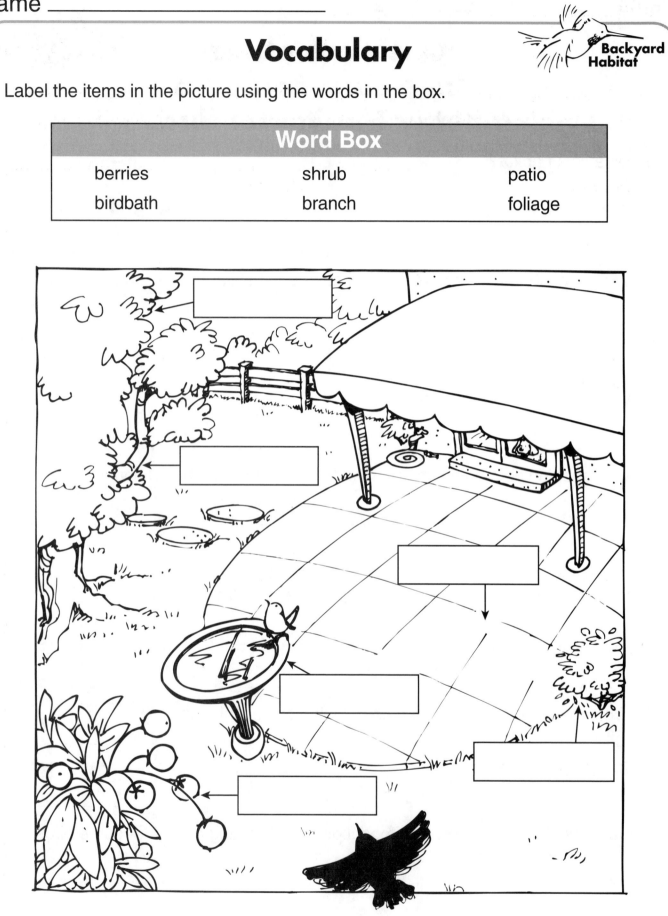

Name _____

Be a Birdwatcher

Backyard Habitat

Be a birdwatcher for one week. Spend time every day looking for birds near your home. Write the name of each kind of bird you see. If you do not know the name of the bird, draw a picture.

Sunday	Monday
Tuesday	**Wednesday**
Thursday	**Friday**
Saturday	

States of Matter

solid

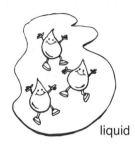
liquid

gas

What Is Matter?

Matter is the material that makes up everything on Earth. Rocks, paper, wood, water, and even air are made of **matter**! Animals, plants, and people are made of matter, too.

Matter is made up of very tiny particles called **atoms**. You can't see these tiny particles. They are far too small. But millions and millions of atoms make up your body, this book, and your pencil. The milk you drink for lunch, your lunch tray, and your fork are also made of millions of atoms.

In some kinds of matter, all the atoms are the same. Matter that is made of all one kind of atom is called an **element**. There are over 100 known elements on Earth. Oxygen is one kind of element. Gold is another. All matter is made from these elements. Most matter, however, is made of a combination of elements. A combination of elements is called a **compound**. For example, when you mix together one part of the element sodium with one part of the element chloride, you get table salt. When you combine two parts of the element hydrogen with one part of the element oxygen, you get water. Compounds are made up of groups of atoms called **molecules**. A single drop of water contains millions of water molecules!

The States of Matter

Matter exists in one of three basic forms: **solid**, **liquid**, or **gas**. These are called the states of matter. The state of a substance, or material, depends on the behavior of its molecules. The same substance can exist as a solid, a liquid, or a gas, depending on the arrangement of its molecules.

Solid

If the molecules that make up a substance are close together and pull on each other with a lot of force, the substance keeps its shape. The molecules don't move around very much at all. This kind of matter is called a solid. A block of wood is solid. A bowling ball is solid. A solid has a definite size and shape. Ice is an example of water in its solid state.

Liquid

In a liquid, the molecules are farther apart. They do not pull so tightly together. The molecules in a liquid move around more freely. A liquid has a certain size or volume, but it does not keep its shape. Instead, it forms itself to the shape of any container you put it in. A liquid like water can pour or flow.

Gas

The molecules in a gas are far apart. They move around easily, without pulling on each other very much at all. Gas has no shape or size of its own. It is sometimes hard to believe that gas is actually matter, but it really is. Take a deep breath. Feel the pressure of the air you have just inhaled as it fills your lungs. Or blow up a balloon and see how the gas presses on the insides of the balloon, making it expand. Water vapor that evaporates from a puddle on a hot summer day is an example of water in its gas state.

Water in Different States

The same substance can exist as a solid, a liquid, or a gas. Water is a kind of matter that can change from one state to another. You can observe these changes. Take an ice cube out of the freezer and set in on a saucer. Check it every few minutes. You will notice the solid turning to a liquid right before your eyes. Leave the saucer on the table for a day or two. You will notice that the water gradually disappears. It is changing into a gas and mixing with the air. You cannot see it, but it is still there.

Solid
The molecules in a solid are very close together.

Liquid
The molecules in a liquid are not as close together.

Gas
The molecules in a gas are very far apart.

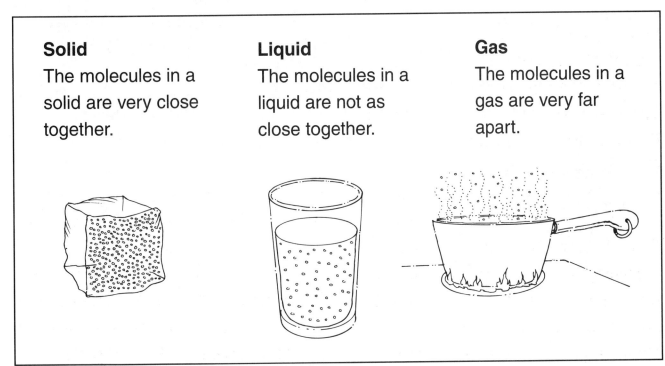

Name _____

Questions about
States of Matter

1. Matter is made up of tiny particles called _____.

   ○ elements
   ○ atoms
   ○ solids
   ○ liquids

2. Matter that is made of only one kind of atom is called _____.

   ○ a compound
   ○ a mixture
   ○ an element
   ○ a block

3. Which one of these is **not** one of the states of matter?

   ○ gas
   ○ solid
   ○ liquid
   ○ heavy

4. Molecules in a gas _____.

   ○ are tightly packed together
   ○ pull on each other with great force
   ○ move around freely
   ○ all of the above

5. Ice is an example of water in which state of matter?

   ○ liquid
   ○ gas
   ○ cold
   ○ solid

Vocabulary

A. Write each word on the line in front of its meaning.

Word Box			
particle	evaporate	atoms	molecule
element	matter	compound	

1. _____ a particular combination of atoms that form a compound

2. _____ a substance whose atoms are all the same kind

3. _____ what all things are made of

4. _____ a combination of elements

5. _____ a tiny piece

6. _____ to turn from liquid to gas

7. _____ microscopic particles of matter that are smaller than molecules

B. Draw a picture to show the meaning of each of these words.

solid	liquid	gas

Name _____

An Experiment to Try

Materials

- 3 ice cubes the same size
- 3 bowls

Procedure

1. Place an ice cube in each bowl.
2. Place one bowl in the sun.
3. Place one bowl in the shade.
4. With the help of an adult, place one bowl in the microwave. Microwave on high for 1 minute. Ask the adult to remove the bowl from the microwave.

Observations

Describe the changes in each ice cube.

	After 1 minute	After 15 minutes	After 30 minutes
Sun			
Shade			
Microwave			

Conclusion

Based on your observations, explain the role of heat in changing water from a solid to a liquid state.

_____

_____

James B. Eads and His Famous Bridge

James Eads (Eedz) was born in Indiana in the year 1820. His family moved around a lot, so James did not get very much schooling. Finally, the family settled in St. Louis, Missouri, on the banks of the Mississippi River. There, James got a job in a store. The storekeeper had a lot of books. James was curious, and he liked to read. The storekeeper let James read his books. James learned a lot. His head was full of ideas. He liked to figure out how to make things and solve problems.

James used to wander down to the Mississippi River and watch the boats go by. He knew that the boats sometimes sank. These boats carried many goods that had come to rest on the muddy river bottom. Some of this cargo was very valuable. James thought he could find a way to collect these things. So he invented a special boat. The boat was connected to a diving bell.

A diving bell is a small vessel that allows people to work underwater. An air hose connects the bell with a boat on the surface of the water.

James could get inside the bell and go to the river bottom. He was able to pick up all kinds of objects. He brought the lost cargo to the surface and sold it. In just 10 years, James became a wealthy man.

When the Civil War began in 1861, James wanted to help the United States. He offered to build boats that were covered with sheets of iron. With these boats, the Union could control the rivers. James hired 4,000 men and built the boats. The boats played an important role in winning the war.

When the war ended, the country began to grow. Trains were a very important form of transportation. A bridge was needed to carry trains across the Mississippi River. James was asked to build the bridge. He hired many men to help him build the bridge. The work began in 1867.

James planned the bridge with care. It would have three arches. It would be made of steel. James wanted the bridge to be very strong. He would use only the finest steel. He checked each shipment of steel. If it wasn't good enough, James sent it back to the factory.

Building the bridge was a tough job. The men had to dig the mud away from the river bottom until they reached solid rock. This was hard and dangerous work. James built a floating hospital to care for anyone who got hurt on the job. It took seven years to build the bridge. When it was finished in 1874, it was the largest bridge in the world. A railroad track was built on the bridge. Later, a highway was added for automobiles. Thousands and thousands of trains, trucks, and automobiles have crossed the Mississippi River on this amazing bridge.

The people of St. Louis decided to name the bridge after James. It is called the James B. Eads Bridge. If you go to St. Louis, you can see this graceful bridge. Many people think it is one of the most beautiful bridges ever built.

James B. Eads began life as a poor boy. He did not have a chance to go to school. He had to learn everything on his own. But he became one of America's foremost engineers. He was a successful inventor. He was a creative thinker. He was able to figure out how to build almost anything he could imagine. The James B. Eads Bridge is a great reminder of this remarkable man.

Questions about
James B. Eads and His Famous Bridge

1. When James was young, he got a job working in a _____.

   ○ bank
   ○ mine
   ○ school
   ○ store

2. James realized that there were valuable things _____.

   ○ on the river bottom
   ○ in the woods
   ○ in Cincinnati
   ○ buried in his backyard

3. James invented ironclad boats to help the United States win _____.

   ○ World War I
   ○ the Civil War
   ○ the War of 1812
   ○ the Revolutionary War

4. In 1867 the people decided to build a bridge across the Mississippi River _____.

   ○ because they thought it would look pretty
   ○ so that automobiles could get across
   ○ so that trains could get across
   ○ because they were tired of riding in boats

5. The bridge was made of steel because steel is _____.

   ○ shiny
   ○ strong
   ○ cheap
   ○ soft

Name _____

Vocabulary

Write the word from the box that means about the same thing as the underlined words in each sentence.

Word Box
steel vessel cargo engineer transportation foremost arches successful

1. It took all day to load the goods and materials onto the airplane.

   _____

2. Albert Einstein was perhaps the most important scientist of his time.

   _____

3. The autoworkers used a strong, hard metal to make the frame of the car.

   _____

4. There were beautiful curved openings in the wall around the garden.

   _____

5. Mrs. Walton planned to use a bus for a way of traveling and carrying things when she took her class to Washington, D.C.

   _____

6. A submarine is a boat that can operate underwater.

   _____

7. When he goes to college, Mark plans to study to be a person who designs and builds things.

   _____

8. After many tries, our experiment was finally turning out right.

   _____

Be a Bridge Builder

1. Build two equal stacks of books or blocks.

   • Make the stacks about 6 inches (15 cm) tall.

   • Place the two stacks on a table, about 10 inches (25.5 cm) apart.

2. Using only **one** sheet of plain copier paper, build a bridge between the two stacks that is strong enough to support the weight of a pen or pencil. You may **not** use tape, staples, or any other materials. Use words and pictures to explain your experiment.

This is what I did.	This is what happened.

3. Try to improve on your design.

This is what I tried next.	This is what happened.

©2002 by Evan-Moor Corp.   Read and Understand, Science • Grades 3–4 • EMC 3304

Turned to Stone

You probably see rocks every day. Rocks are not alive. But did you know that one type of rock is made from parts of creatures that were once alive? That rock is called limestone.

Limestone is a type of rock called sedimentary rock. Most limestone is formed from the shells of tiny animals that live in water. Limestone is also formed from the skeletons of an animal called coral. When the animals die, their shells float to the bottom of the ocean and break into many pieces. Over thousands of years, the layers of shell pieces build up. Upper layers press down on lower layers. The layers of shell pieces turn into solid rock. This rock is limestone. Sometimes you can see tiny pieces of animal shells in limestone.

Limestone is very soft compared to most rocks. Flowing water can easily wear away pieces of limestone. That is why many caves are made out of limestone.

Sometimes nature carves limestone into interesting shapes. In China there is a beautiful rock formation. Over thousands of years, rain dissolved a huge block of limestone. Harder rocks within the limestone did not dissolve. These rocks can still be seen. They look like eggs sticking out of the ground.

If you find a rock that you think is made of limestone, there is an easy way to check. Fill an eyedropper with strong vinegar. Then squeeze a few drops onto the rock. Watch the rock through a magnifying glass. If it is limestone, the surface of the rock will fizz and bubble. The vinegar reacts with the limestone to produce a gas.

Limestone is often used for building. The rock is dug out of the ground in places called quarries. Limestone is cut into large blocks. These blocks can be used to build churches or other large buildings. In Europe, many old buildings are made of limestone. Because the rock is soft, it can be carved. You

©2002 by Evan-Moor Corp.

Read and Understand, Science • Grades 3–4 • EMC 3304

can often see these carvings in old churches, palaces, and other buildings. Limestone is also crushed into tiny stones. This crushed stone is used to build roads and sidewalks.

Ancient people also used limestone to make art. In England there are many limestone cliffs. People carved pictures of horses and other things into the limestone. These huge carvings can still be seen today.

Limestone truly has an amazing story. It is formed from parts of once-living animals in the ocean. These parts are changed into a special kind of rock. People might use that rock to create a work of art or the pavement under your feet. That's a long way to travel from the bottom of the sea!

Rocks Times Three

There are three types of rocks:

Sedimentary *rocks are formed when particles of matter are laid down in many layers and pressed together.*

Photo by Nick Ferris

Igneous *rocks are formed when hot liquid rock from the center of the Earth comes to the surface, cools, and hardens.*

Photo by Nick Ferris

Metamorphic *rocks are formed when heat and pressure inside the Earth change one kind of rock into another kind of rock.*

Photo by Nick Ferris

Name _____

Questions about *Turned to Stone*

1. How is limestone formed?

   _____

   _____

   _____

   _____

2. What type of rock is limestone?

   _____

3. Why are there likely to be caves in limestone formations?

   _____

   _____

4. What is one way to tell whether a rock is limestone?

   _____

   _____

5. Name three ways people use limestone.

   _____

   _____

6. You are on a hike and you find a limestone cliff. What do you know about what that land was like millions of years ago? Support your answer with information from the story.

   _____

   _____

   _____

Vocabulary

Use the words in the box to help you unscramble each numbered word. Then find the correct meaning below and write the letter on the short line.

Word Box				
coral	dissolves	quarries	igneous	limestone
sedimentary	metamorphic	ancient	skeletons	magnifying

1. seguino   _____   _____

2. arcmeotimhp   _____   _____

3. lorca   _____   _____

4. reusqira   _____   _____

5. leteksnso   _____   _____

6. rinsdaeyetm   _____   _____

7. svisodels   _____   _____

8. gfymiignna   _____   _____

9. stmileeno   _____   _____

10. tneicna   _____   _____

a. causing to appear larger
b. a type of animal that lives underwater
c. supporting bony structures of vertebrates
d. disappears when mixed with water
e. places where rocks are dug out of the ground
f. any rock formed by hot liquid from the center of the Earth
g. belonging to times long ago
h. any rock formed when heat and pressure change one type of rock into another
i. any rock formed when layers of matter are pressed together
j. a sedimentary rock formed from shells and skeletons of tiny sea animals

Table of Contents

Here is a table of contents from a book about rocks. Study it and then answer the questions below.

1. Which comes first, the chapter on igneous rocks or the chapter on sedimentary rocks?

   _____

2. On what page does Chapter 4 begin?

   _____

3. Which chapter would give you information about limestone?

   _____

4. Which chapter would tell you about storing a rock collection?

   _____

5. If you don't know the meaning of a word you read in this book, in which section would you look?

   _____

6. In which chapter might you be able to read about carving statues from stone?

   _____

Always Pointing North

If you hold a compass in your hand, the needle will always point toward the North Pole. This is true no matter where you are on Earth. Whether you're on a ship in the middle of the Pacific Ocean, walking down a busy street in Paris, or standing on top of Mount Everest, your compass will still point north.

The most common type of compass is the **magnetic compass**. A magnetic compass works because the Earth has a **magnetic field**. Scientists think the red-hot, molten iron in the middle of the Earth's center, or core, causes this magnetic field. As the Earth spins, this liquid iron spins, too. Scientists think that the spinning liquid iron creates a weak magnetic force. This happens because iron is a magnetic material.

So why does a compass needle always point north? First you need to know that a compass needle is really a magnet. Every magnet has a south pole and a north pole.

Now, imagine that the spinning liquid inside the Earth is like a huge, buried magnet. The south end of the magnet rests on the North Pole. The north end rests on the South Pole. When it comes to magnets, opposites attract. That means that the south end of one magnet will attract the north end of another magnet. So the North Pole (which has the south end of the Earth's magnet) attracts the north end of the compass needle. That's why the needle always swings around to point north.

North Pole

South Pole

The Earth's magnetic field is very weak, so it can only move small objects. For this reason the compass needle has to be very lightweight. It also needs to be able to spin easily without **friction** slowing it down.

Along with the needle, a compass also includes a **compass card**. This card is marked with four **cardinal points** that show the four major directions— north, east, south, and west. Between each cardinal point is an **intercardinal point**—northeast, southeast, southwest, and northwest. A compass card may also be marked with the 360 degrees that make up a circle. These markings help the person using the compass figure out the exact direction of travel.

People have been using compasses for almost a thousand years. Chinese **navigators** used them to guide their ships during the 1100s. These early compasses were very simple. They were just a piece of magnetic iron floating on a cork in a bowl of water. Still, the navigators would have been lost without them! By using a compass, these navigators could figure out which way to head, even if they were in the middle of the ocean, or if the sun had disappeared behind the clouds.

So if you want to make sure you're not lost—or if you just want to have some fun—carry a compass in your pocket. It will always point you in the right direction.

Make Your Own Magnetic Compass

You can make a simple compass like the ones the ancient Chinese navigators used. You will need:

- *a sewing needle*
- *a bar magnet*
- *a cork or the cap from a plastic milk jug*
- *a bowl of water*

1. *Run one end of the magnet along the needle. Do this about 20 to 30 times. Be sure you always rub the magnet along the needle in the same direction.*

2. *Put the cork or milk cap in the middle of your bowl of water.*

3. *Lay your magnetic needle on top of the cork or cap. The needle should spin around slowly until it is pointing north. Check your homemade compass against a real compass to see if both needles point in the same direction.*

Questions about
Always Pointing North

1. The Earth's core is filled with _____.

   ○ magnets
   ○ molten iron
   ○ rocks
   ○ nothing

2. A compass needle is really a _____.

   ○ sewing needle
   ○ wire
   ○ toy
   ○ magnet

3. All magnets have _____.

   ○ a compass
   ○ a south pole and a north pole
   ○ iron
   ○ an east pole and a west pole

4. The first people to use compasses were _____.

   ○ Chinese navigators
   ○ American astronauts
   ○ pirates
   ○ British explorers

5. Compasses are good to have because _____.

   ○ they are easy to make
   ○ they always point south
   ○ they help us find our way
   ○ they are magnetic

6. Compass needles must be lightweight because _____.

   ○ the Earth has a weak magnetic field
   ○ they are expensive
   ○ there isn't much metal to be used
   ○ compasses are small

Name _____

Vocabulary

A. Match each word with its meaning.

magnetic   the center part of something

molten   north, south, east, west

core   an instrument for telling direction

friction   attracting iron or steel

navigator   melted by heat

compass   northeast, southeast, northwest, southwest

cardinal points   rubbing together

intercardinal points   someone who guides a ship

B. Use the best word from the list above to complete these sentences.

1. The _____ piloted the ship.

2. The road sign directed us north, which is one of the _____

   _____.

3. The iron filings stuck to the bar because the bar was _____.

4. Were it not for our _____, we would have stayed lost
   in the woods.

5. There was a lot of _____ when the two
   objects rubbed together.

6. Traveling from Iowa to Florida, we drove southeast, which is one of the

   _____ _____.

7. _____ lava poured out of the volcano.

8. The center of the Earth is called the _____.

Name _____

Read a Map

Pretend you are going on a hike in the park. Of course, you bring your compass along to help you find your way! You also have the map below to help you. Look at the map. Then use the information on it to answer the questions.

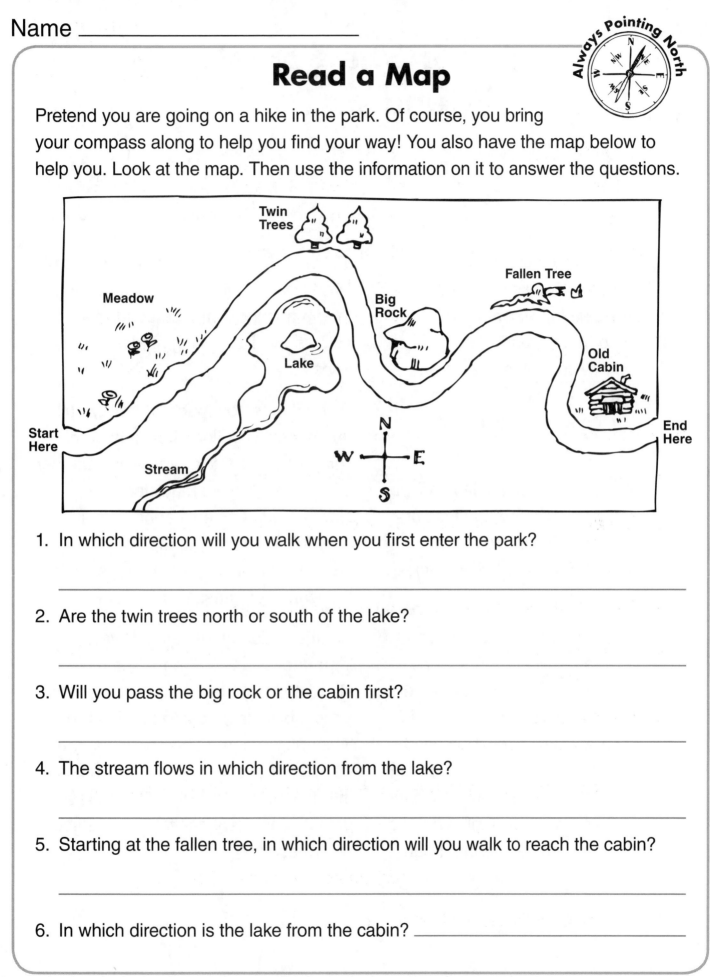

1. In which direction will you walk when you first enter the park?

_____

2. Are the twin trees north or south of the lake?

_____

3. Will you pass the big rock or the cabin first?

_____

4. The stream flows in which direction from the lake?

_____

5. Starting at the fallen tree, in which direction will you walk to reach the cabin?

_____

6. In which direction is the lake from the cabin? _____

Fire in the Forest: Friend or Foe?

"Only YOU can prevent forest fires!" Smokey the Bear has been telling us to be careful with matches for many years. And it is still good advice. Forest fires are dangerous. They can burn down homes, harm animals, and destroy trees.

But forest fires are not all bad. In fact, ecologists have learned that fires are a natural part of life in the forest. They know that fire sometimes helps a forest stay healthy.

How does fire help? As you know, a forest is an area where many trees grow. Over time some of the trees die. Branches fall to the ground. Leaves collect on the forest floor. This dead material is called litter. Fire can clear the litter away, leaving more space for the trees to grow. If the litter is not too deep, the fire burns along the ground.

It stays low. It may scorch the trunks of large trees, but it cannot reach up into their branches. The burned litter also acts as fertilizer, giving nutrients to the trees.

Without fires, too much litter may build up on the forest floor. It may become very deep and thick. A careless camper or a lightning strike may start a fire. Now, the fire fighters cannot stop it. The fire is too big and too hot. The fire blazes high into the treetops, killing thousands of trees.

But these dead trees are still very important. Many of them remain standing for years. They are called snags. Insects bore into the dead wood. Woodpeckers and other birds eat the insects. These birds also make nests in the dead trees. In fact, burned forests provide important habitat for a number of bird species.

Fire helps the forest in another way. Many forest plants have seeds that can only sprout if they are heated by fire. Lodgepole pinecones only open when exposed to fire. Other seeds have tough coatings that must be burned away. Many non-native species of plants are killed by fire. These are plants that do not belong in the forest. Fire helps to maintain a healthy community of forest plants.

Many fires occur naturally in the forest. Most of them are started by lightning strikes. They usually happen in the summer when the weather is hot and dry. For many years, fire fighters raced to put out every forest fire. But now, many fires are allowed to burn. Fire fighters try to protect houses and other buildings. People who live in or near the forest must understand that fire might be a threat.

Everyone should follow Smokey's rules. No one should ever play with matches. When we go out in the woods, we should make sure to put our campfires out before we leave our campsites. Fire is serious and can be deadly. We still need Smokey to remind us to be careful with fire. But we also know that fire can sometimes be a friend.

A Real Live Smokey

In 1944 Smokey the Bear became the symbol for fire prevention seen on posters and in advertising. Then in 1950 a black bear cub was found clinging to a tree after a fire in the Lincoln National Forest in New Mexico. After his wounds were healed, the little bear, by now named Smokey, was sent to live at the National Zoo in Washington, D.C. Over the years, thousands of people from all over the world came to see this living symbol of fire prevention. When Smokey died in 1976, he was buried beneath a stone marker in Smokey Bear State Park in Capitan, New Mexico.

Name _____

Questions about
Fire in the Forest: Friend or Foe?

1. List three ways that forest fires are harmful.

   _____

   _____

   _____

2. List three ways that fire helps the forest.

   _____

   _____

   _____

3. Name two ways to help prevent forest fires.

   _____

   _____

4. How are most natural forest fires started?

   _____

5. Why do you think forest fires happen mostly in the summer?

   _____

6. Do you think people should be allowed to build houses in the forest? Explain your answer.

   _____

   _____

Vocabulary

Write the word from the box that means about the same thing as the underlined word or words in each sentence.

Word Box				
habitat	forest	an ecologist	species	occur
snags	litter	prevent	nutrients	community

1. The scout troop went for a six-mile hike through the <u>woods</u>.

   _____

2. The <u>layer of leaves and twigs</u> beneath the trees made a soft seat for them to rest on when they became tired.

   _____

3. The leader of the hike was <u>a person who studies how living things relate to each other</u>.

   _____

4. "The trees depend on <u>things needed for good health</u> from the soil, and the forest <u>animals living in the same area</u> depend on the trees for food and shelter," she explained.

   _____

   _____

5. She explained that the area around a small creek was a good <u>place to live</u> for frogs.

   _____

6. She showed the scouts some <u>tall, dead trees that were still standing</u>.

   _____

7. "Even though people try to <u>stop</u> fires, they still <u>happen</u>," she explained.

   _____

   _____

8. Together they counted eight different <u>kinds</u> of birds on their hike.

   _____

Name _____

Life Cycle of a Lodgepole Pine

Read about the events in the life cycle of a lodgepole pine. Number them in the correct order.

©David R. Bridge

_____ Lodgepole pinecones open only when they are exposed to fire. The seeds are then released from the cone.

_____ After growing for many years, the lodgepole pine tree is now mature.

_____ When conditions are right, the lodgepole pine continues to grow. The tiny tree will have plenty of room because the fire has removed litter from the forest floor.

_____ A lightning strike starts a forest fire. The flames kill the lodgepole pine.

_____ The lodgepole pine seeds sprout when there is sufficient moisture and soil in which the roots can grow.

The Story of Oil

Millions of years ago, most of the Earth's surface was covered with water. Giant oceans spread across the globe. Tiny plants and animals floated in these oceans. Other plants and animals lived along the shorelines.

As the plants and animals died, their bodies sank to the bottom of the ocean. Over millions of years, these remains formed deep layers on the ocean floor. As new layers were added, the layers on the bottom were squeezed and pressed. The weight of the water added more pressure. Bits of sand and mud sank to the bottom, too. These added even more pressure. Gradually, the bits of mud and sand were squeezed into solid rock. The dead plant and animal material was slowly changed into a material we call oil.

Oil is also called **petroleum**. This word comes from two Latin words that mean "rock oil." That's because oil lies trapped between layers of rock deep in the Earth. Large drilling machines punch holes through layers of soil and rock to find the oil. The oil is then pumped to the surface. It is piped into large tanks. It is then sent to a special factory called a refinery. At the **refinery**, the oil is separated into useful products.

Much of the oil is made into **fuel**. Fuel is any material that is burned to produce heat or power. Several different kinds of fuel can be made from oil. **Gasoline** is a fuel that your family probably uses every day. Your car's engine burns gasoline to get the power it needs to run. Heating oil and natural gas are fuels we use for cooking and to heat our homes. Jet airplanes run on jet fuel, which is also made from oil.

Hundreds of other important products are made from oil. **Asphalt**, the material we use to pave roads, is made from oil. So are **plastics**, some **medicines**, and makeup, just to name a few. It is easy to see that we rely on oil for many things.

But there are two big problems with using oil. The first is that we are steadily using the oil that lies under the ground of our planet. It takes millions of years for oil to form, and we are using it up much faster than it can be replaced. The second problem is that the fuels made from oil cause a lot of **pollution**. When we burn these fuels, they produce a lot of unwanted gases and soot. We need to find other ways to heat our homes and power our cars.

Scientists are working on some **alternative** kinds of fuel. They have learned to use the energy from the sun and wind to make power. They have discovered ways to make fuel from corn and other plants. These fuels do not produce as much pollution, and they are **renewable**, meaning we can make more of them. Hopefully, it won't be long before we can all use these cleaner sources of energy.

How You Can Help

Everyone can help make things better simply by using less fuel. Here's how:

- *Drive cars that can travel many miles on one gallon of gas.*

- *Turn down the heat in your home and wear a sweater.*

- *Walk to the store, or take the bus.*

- *Ride a bike to school or work, or join a carpool.*

Solar panels can heat water for a swimming pool.

These windmills are used to generate electricity.

Name _____

Questions about
The Story of Oil

1. Oil is formed from _____.

   ○ the remains of ancient plants and animals

   ○ sand and gravel

   ○ water

   ○ mud

2. Another word for oil is _____.

   ○ energy

   ○ petroleum

   ○ medicine

   ○ pollution

3. Oil can be used to make _____.

   ○ gasoline

   ○ plastic

   ○ asphalt

   ○ all of the above

4. One of the main problems with using oil is that _____.

   ○ it is too expensive

   ○ we are using it all up

   ○ it doesn't work very well

   ○ we can't find it

5. Scientists have found ways to use the energy from _____.

   ○ the sun

   ○ wind

   ○ corn and other plants

   ○ all of the above

Name _____

Vocabulary

A. Write a **T** in front of each statement that is true.

Write an **F** in front of each statement that is false.

1. _____ Any material that is burned to produce heat or power is called fuel.

2. _____ A refinery is a factory where oil is made into useful products.

3. _____ The word petroleum comes from two Latin words meaning "dead plant."

4. _____ Gasoline is the fuel that powers most cars.

5. _____ Pollution occurs when we dirty the air, land, or water.

6. _____ Asphalt is a material that is used to pave roads.

7. _____ **Soil** is another name for plastic.

8. _____ **Squeeze** means to press very hard.

9. _____ If a resource is renewable, we don't have much of it left.

10. _____ If you have an alternative, you have no other choice.

B. Energy provides the power to make things work. List three sources of energy that are alternatives to oil.

_____

_____

_____

C. When you are 40 or 50 years old, what do you think will power cars? Explain why you think that.

_____

_____

_____

_____

Name _____

Early Uses for Oil

In some places, oil seeps to the surface of the Earth and forms sticky pools. Long ago, people found these pools. They did not know how to make gasoline or plastic from petroleum. But they did find uses for the thick, gooey oil that they found in these **pools**. This sticky substance is called **pitch**. According to the Bible, Noah used pitch when he built the Ark. The Egyptians used pitch to coat **mummies**. It was also used to pave the **streets** of ancient cities.

In this country, Native Americans smeared pitch over the outside of **canoes** and **baskets** to make them waterproof. Later, American colonists found that this substance made good axle **grease** for their **wagons**. And Spanish settlers spread it over the roofs of their **adobe houses** to keep out the rain.

A. Unscramble these words from the passage above.

immusem _____   skasbte _____

chipt _____   lopos _____

snocea _____   ushsoe _____

snagow _____   eerttss _____

rgeesa _____   obead _____

B. Choose two of the words that you unscrambled and write a sentence with each.

1. _____

_____

2. _____

_____

Sally Fox: Spinning a Life

Photo used by permission of Sally Fox.

Do you know the story of Rumpelstiltskin? It's the story of a funny little man who could spin straw into gold. Sally Fox must have loved that story when she was a little girl. You see, Sally Fox loves to spin. When she was 13 years old, her parents gave her a real spinning wheel. She would collect cotton from pill bottles. She would spin the cotton into string. She would brush her fuzzy dog and spin his hair into yarn. Sally spent hours at her spinning wheel.

As Sally grew up, she found other interests. But she never stopped spinning. When she went to college, she studied biology. She wanted to learn everything she could about the natural world. She wanted to help protect the environment.

Sally decided to study insects. She knew that some insects eat crops. Farmers spray pesticides on the crops to kill the insects. Sally did not like pesticides. She knew they could harm many living things. She wanted to figure out how to raise plants that did not need pesticides. She got a job working with a plant breeder. The job was to breed cotton plants that could resist insects.

One day while she was at work, Sally found some strange seeds. These seeds grew cotton that was brown. Most farmers grow white cotton. They don't like brown cotton because the fibers are short. It is hard for big machines to spin. But Sally knew that brown cotton could be grown without pesticides. And she liked the idea of cotton that had a natural color. So she planted some of the seeds.

Sally was excited about the brown cotton that grew from her seeds. She began to work with the plants. She saved seeds from the plants with the longest fibers and deepest colors and replanted them. It took many years, but finally Sally was able to grow colorful cotton with long fibers. Some of the cotton was a beautiful reddish brown. Some was a deep chocolate color. There were even three different shades of green cotton.

People liked this new cotton. It could be spun by machine. It did not need any dyes. It did not contain any pesticides. Soon, Sally's cotton was being used to make clothing, sheets, and towels. Sally could not grow enough cotton by herself. She had to get other farmers to grow her cotton, too.

Sally still works with cotton. She runs a business called Natural Cotton Colours, Inc. Sally grows thousands of cotton plants every year. She walks through the fields and inspects each plant. She measures the length of the fibers. She looks carefully at the color. She takes notes about each plant. She saves the seeds from the best plants. Sally's cotton keeps getting better and better.

Many organizations have given Sally awards. She has won awards for helping the environment. She has also won awards for being an inventor.

Perhaps Rumpelstiltskin really was able to spin straw into gold. But Sally Fox has done something even more amazing. She has found a way to spin cotton into a lifetime of achievement.

The Cotton Plant

Cotton fibers come from the fruit of the cotton plant, called the boll. When the boll opens, you can see the fluffy fiber that has been growing inside. There are also seeds in the boll. These must be separated from the fiber. The seeds can be crushed to make cottonseed oil, used in food products such as salad dressings and margarine.

Name _____

Questions about
Sally Fox: Spinning a Life

1. Why did Sally decide to study biology in college?

   _____

   _____

   _____

2. Naturally colored cotton has shorter fibers than white cotton. Why was this a problem?

   _____

   _____

3. What are some items that are made of cotton?

   _____

   _____

   _____

4. Why does Sally spend time looking at each cotton plant she grows?

   _____

   _____

   _____

5. What are the advantages of the cotton Sally has produced?

   _____

   _____

   _____

Name _____

Vocabulary

A. Write the number of each word on the line in front of its meaning.

1. pesticide

2. fiber

3. dye

4. inspect

5. spin

6. organizations

7. biology

8. resist

9. achievement

10. breed

_____ groups who work for a certain goal

_____ to plan the reproduction of plants or animals

_____ to look at very carefully

_____ a chemical used to kill insects

_____ to not be affected by; stand up against

_____ to twist fibers into thread or yarn

_____ an accomplishment; success

_____ a thin strand of material

_____ the study of living creatures

_____ a substance used to color cloth

B. Draw a picture of three things that you have at home that are made of cotton. Label each picture.

Name _____

Sally Fox

Spinning a Tale

Tell the story of Rumpelstiltskin in your own words. If you do not know the story, make up a story of your own that includes a spinning wheel. Use additional paper if you need more room.

_____

_____

_____

_____

_____

_____

_____

_____

_____

_____

_____

_____

_____

_____

_____

_____

_____

Keeping Warm for Winter Fun

Ricky loves snowboarding. Almost every weekend, he goes to the ski area near his home in Vermont. He is happiest when there is fresh, new snow to enjoy. He doesn't seem to mind the cold at all. That's because he has plenty of insulation.

What is insulation? Insulation keeps heat from moving from one place to another. Ricky wants to keep the heat his body generates near his skin. So before he hits the slopes, Ricky prepares very carefully. He makes sure to dress in layers. He puts on long underwear and socks that are made of special fibers. These fibers contain lots of tiny air pockets. The air pockets trap warmth and keep it next to Ricky's body. He puts on lined pants and a warm jacket. He wraps a scarf around his neck. He puts on a soft, fluffy hat. Finally, Ricky pulls on thick gloves. He's ready to go.

Ricky hops onto the chairlift. Up into the air, higher and higher, the chair climbs. Ricky can't wait to join the skiers and snowboarders gliding down the trails below. When his chair reaches the summit, he leaps off. He pauses for a moment at the top of the slope, then pushes off. Down the mountain he flies, carving through the fresh powder, plumes of snow spraying behind him. When he reaches the bottom, Ricky heads straight for the chairlift again.

After a few trips down the mountain, Ricky decides to take a break. Even though his clothes give him lots of insulation, he begins to get cold as he takes off his equipment and walks toward the lodge. He knows the lodge will be warm. It is so warm in fact, that Ricky pulls off his jacket and hat as soon as he walks through the door.

How does the lodge stay so warm on a cold winter day? A furnace in the basement of the lodge burns heating oil. Warm air is blown out of the furnace to all the rooms in the lodge. The workers who built the lodge used insulation to help keep the heat inside the building. They put thick layers of cottony fiberglass in the walls. Because heat rises, they put even thicker layers in the attic.

©2002 by Evan-Moor Corp.

Read and Understand, Science • Grades 3–4 • EMC 3304

Ricky and his friends enjoy a cup of hot chocolate, then they head back out onto the slopes. As they ride the chairlift up to the top of the mountain, they notice several little animals. They see chickadees flitting in the trees and a rabbit bounding across the slope. They wonder how the animals manage to stay warm in the cold weather.

Animals have their own ways of keeping warm. When birds are cold they fluff up their feathers. This creates many small pockets of warm air between the feathers. (Like the air pockets in Ricky's clothing!) Rabbits and other mammals grow extra fur for the winter. They often eat extra food to build a layer of fat as winter approaches. A thick fur coat makes good insulation. So does a thick layer of fat. Both act in the same way that a good jacket does. They prevent the body's heat from escaping.

The animals, in their own winter "snowsuits," are soon out of sight. Ricky and his friends are already talking about which trail they will try on the next run.

The sun sinks low in the west, and all too soon the day is ending. Ricky flies down the mountain one last time. Then he hurries to the bus that is waiting in the parking lot. As the bus pulls away, fat snowflakes begin to fall. Ricky closes his eyes and smiles.

Animals Keep Warm in Winter

Animals have various ways to stay warm during the cold winter months. Some, like the buffalo in the photograph, grow a heavy winter coat. Birds fluff up their feathers to trap an insulating layer of air next to their skin. Hibernating animals, such as bears, eat a lot of food to build up an insulating layer of fat.

Digital Stock

Questions about
Keeping Warm for Winter Fun

1. What does insulation do?

_____

_____

2. Why does Ricky cover most of his body with clothing?

_____

_____

3. How do layers of clothing help Ricky stay warm?

_____

_____

4. What are two kinds of insulation found on animals' bodies?

_____

_____

5. How many ways can you think of that your house is kept warm? List the ways. Mark an **X** by each way that uses insulation.

_____

_____

_____

_____

_____

Find the Right Word

Choose a word from the box to complete each sentence.

Word Box					
insulation	fibers	plumes	lodge	furnace	powder
chickadee	flitting	bound	chairlift	summit	slope

1. Our family stayed at a _____ in the mountains.

2. A _____ ate seeds at the bird feeder.

3. The graceful deer can jump the fence in a single _____.

4. Megan wrapped a layer of _____ around the water pipe to keep it from freezing.

5. Two people can ride together on the _____ to the top of the ski run.

6. The house was cold, so Grandpa turned on the _____.

7. Amie sat in the grass and watched the butterflies _____ from flower to flower.

8. When they reached the _____ of the mountain, the climbers felt a great sense of accomplishment.

9. The _____ in the wool blanket were soft and fuzzy.

10. Aunt Rachel wore a hat that was decorated with fluffy ostrich feather

_____.

11. Dad gave the baby a bath and then sprinkled her with _____.

12. I have never skied down such a steep _____!

78

Name _____

Insulation in Our Daily Lives

We use insulation in many ways. We carry hot drinks in Thermos™ bottles. These bottles have a thick layer of insulation around the outside. This layer of plastic or glass helps prevent heat from moving from the hot chocolate inside the Thermos™ to the cold air outside.

Insulation can also help keep things cool. We pack soft drinks and sandwiches in a cooler to take to the beach. We want the things in the cooler to stay cool. The cooler is made of a plastic material that slows the movement of heat from outside the cooler to inside the cooler.

The objects pictured below provide insulation. Write about what each object is used for and how it provides insulation.

_____
_____
_____
_____

_____
_____
_____

_____
_____
_____

Marc Hauser: Learning About Animal Minds

Do you have a dog, cat, or parakeet for a pet? Do you ever wonder what your pet is thinking when it looks at you? Do you wonder whether animals feel sad or happy the way you do? It sometimes seems that they do, especially dogs with their sorrowful eyes and wagging tails. But how can we be sure?

Marc Hauser is a scientist. He works with animals. He wants to understand how animals think. He watches animals in the wild to learn about their lives. He watches to see what kinds of problems the animals face. He wants to find out how they solve those problems. Then he sets up experiments in the laboratory. These experiments help him learn what the animals are thinking.

Here is an example. Marc wanted to know what animals understand about tools. Scientists know that animals often use tools. Monkeys and apes in the wild break off branches to get termites out of holes in the ground. They use rocks to crack open hard nuts. They use broad leaves as cover from the rain. But, when animals use these tools, do they really know what they are doing? What would happen if we snuck into the forest and painted a broad leaf pink? Would they not use that leaf? Or would they understand that its color was not important? Humans know that what makes a good tool is not its color. Rather, what makes a tool good is its shape and what it's made of. If something looked like a hammer but was made of soft rubber, it wouldn't be a good hammer. If we painted a real hammer rainbow colors, it would still work well.

In the laboratory, Marc gave cotton-top tamarin monkeys some tools to work with. Some of the tools had been painted odd colors. When the tamarins needed a tool, they often chose the best one for the job. They didn't pay attention to color, or other unimportant features of the tool. They seemed to be able to figure out which tool would work best. Marc's experiments showed that tamarins do have some understanding of how simple tools work.

©2002 by Evan-Moor Corp.

Read and Understand, Science • Grades 3–4 • EMC 3304

Marc is also interested in how animals think about numbers. Do they understand numbers at all? Marc did an experiment to find out. Here's how it worked. In your mind, picture a stage. At the front of the stage is a screen that blocks your view. The person doing the experiment shows you two dolls. Then he sets the dolls behind the screen. Next, he moves the screen away. How many dolls do you expect to see behind the screen? That's right, you expect to see two dolls. You would be surprised to see one doll or three dolls.

Tamarin monkeys live in South America.

Marc did this experiment with rhesus monkeys. Instead of dolls, he showed the monkeys two purple eggplants. He placed the eggplants behind the screen. Sometimes, before he moved the screen, he would secretly add an extra eggplant or take one away. Then he removed the screen. When the number of eggplants behind the screen matched the number the monkeys expected to see, the monkeys looked at the eggplants for a short time, about one second. When the numbers did not match, the monkeys looked for a much longer time, about three to four seconds.

Marc did the experiment again and again. Some other scientists tried the experiment, too. They tried it with other monkeys. Each time the results were the same. When the numbers matched, the monkeys looked for a short time. When the numbers did not match, the monkeys looked much longer at the stage. It seems that the monkeys do have some understanding of numbers.

Marc still has many questions to answer. He hopes to learn more about how animals think and feel. He wants to understand the differences in animal brains. He thinks that learning more about animals may help us to know more about ourselves.

Name _____

Questions about
Marc Hauser:
Learning About Animal Minds

1. Give three examples of the ways monkeys and apes use tools in the wild.

   _____

   _____

   _____

2. What happened when Marc gave the monkeys some tools to use?

   _____

   _____

3. a. Why did Marc do the experiment with the eggplants?

   _____

   _____

   b. What happened?

   _____

   _____

   _____

   c. What do you think this means?

   _____

   _____

   _____

Vocabulary

A. Write each word on the line in front of its meaning.

Word Box			
laboratory	termites	experiment	broad
expected	stage	screen	humans

1. _____ wide

2. _____ a room or building where scientific work takes place

3. _____ a panel used to keep something out of view

4. _____ a platform where plays are often performed

5. _____ a test that is designed to prove or discover something

6. _____ people

7. _____ insects that eat wood

8. _____ planned on

B. Use a word from above to complete each sentence.

1. There are places on Earth where no _____ live.

2. Our neighbors called a pest control company to rid their house of

   _____ .

3. This _____ makes medicines to treat many kinds of illnesses.

4. It took days for the caravan to cross the _____ expanse of desert.

5. I'm surprised that she has not arrived. I _____ her much earlier.

Name _____

Meet the Monkeys

There are many different kinds of monkeys. These are two kinds of monkeys from the story.

The cotton-top tamarin lives in Colombia. This is a very small monkey that is only about 8 inches (20 cm) tall. Tamarins live in groups and make their homes in trees. Tamarins eat fruit and insects.

Rhesus monkeys live in India. They live in groups. They make their homes on the ground. They eat fruit, seeds, leaves, roots, and insects. Rhesus monkeys are about 2 feet (0.6 m) tall.

Use the information in the pictures and the captions to fill in the Venn diagram.

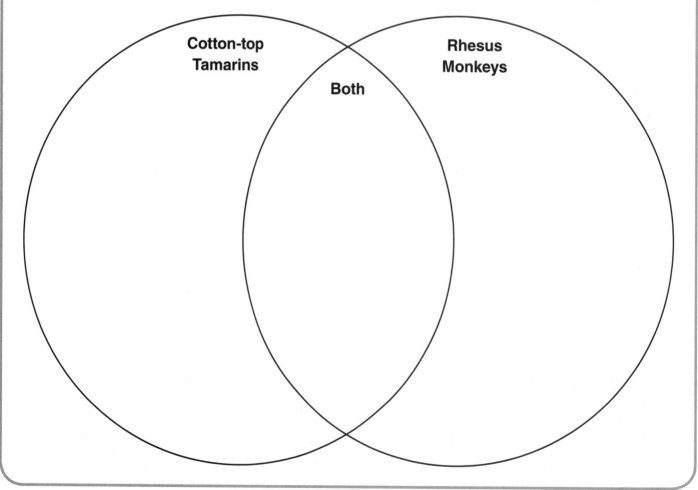

Cotton-top Tamarins

Both

Rhesus Monkeys

©2002 by Evan-Moor Corp.

Read and Understand, Science • Grades 3–4 • EMC 3304

Planetary Almanac: Interesting Facts About Our Solar System

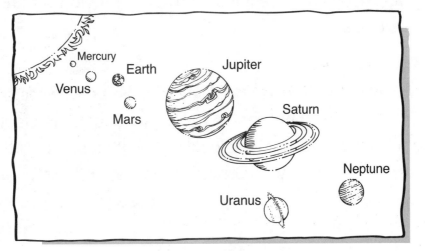

Sun Cycle

Every 11 years, the Sun goes through a cycle. Part of the cycle is calm and quiet. The other part of the cycle is busy and active. During the busy part of the cycle, there are lots of **solar flares**. A solar flare occurs when the Sun spews out high-energy particles. Solar flares can cause problems for satellites. They may even affect Earth's weather. The space probe *Ulysses* is circling the Sun, trying to learn more about the solar cycle.

Memo from Mercury

Mercury is the planet that is closest to the Sun. The surface of this planet can reach 800°F! Mercury is not much bigger than the Earth's moon. It is covered with craters. Scientists think that long ago there were active volcanoes on this planet.

Don't Vacation on Venus

Venus is our closest neighbor in the solar system. But don't plan to take a trip there. You won't like it at all! The atmosphere is filled with clouds of **sulfuric acid**. You won't be able to breathe. And the temperature on Venus is about 900°F. That's too hot for earthlings!

Earth: The Living Planet

Earth is the only planet that is known to support life. The many different plant and animal life-forms on Earth need different kinds of living conditions. The Earth's **biomes** provide these different living conditions. A biome is a region of Earth that has a particular climate that meets the needs of certain plants and animals. A desert biome is very hot and dry. A rainforest biome is warm and wet. Oceans, grasslands, and temperate forests are some other types of biomes.

The Red Planet

Mars is known as the red planet because it is covered with red soil. The latest pictures from Mars show ripples on the surface of the planet. Were these ripples formed by water? That would mean that Mars was once warmer and wetter than it is now. Or were these ripples formed from windblown layers of dust? More exploration is needed to find out the answer to this question.

Jupiter Journal

Jupiter has 17 moons. One of these moons is called Europa. Scientists believe that Europa is covered with ice. They think that an ocean of salt water lies under the ice. If there is water, there might be life-forms on Europa!

Saturn Has the Most Moons

In the year 2000, astronomers discovered 10 new moons around Saturn. They are small moons and are probably made of ice. Saturn now has 28 known moons, more than any other planet.

A Uranus Year

A year is the length of time it takes for a planet to circle all the way around the Sun. A year on Uranus is equal to 84 years on Earth! On Uranus each season lasts more than 20 Earth years. Imagine if winter began when you were 10 years old, and spring didn't come until you were 30!

Far Away Neptune

Neptune is the fourth-largest planet. If it were hollow, it could almost hold 60 Earths! Its size doesn't make it easy to see though. That's because it is also the last planet in our solar system. It is over 2 billion miles from the Earth. It is so far away you can't see it with your eyes. You can barely see it with a telescope!

Name _____

Questions about
Planetary Almanac: Interesting Facts About Our Solar System

1. The Sun goes through a cycle about every _____.

   ○　4 years
   ○ 100 years
   ○　37 years
   ○　11 years

2. The planet that is closest to the Sun is _____.

   ○ Venus
   ○ Mercury
   ○ Earth
   ○ Uranus

3. The planet with the most moons is _____.

   ○ Jupiter
   ○ Mars
   ○ Saturn
   ○ none of the above

4. Our closest neighbor in the solar system is _____.

   ○ Mars
   ○ Venus
   ○ Neptune
   ○ Pluto

5. A year on Uranus is equal to _____.

   ○ 84 Earth years
   ○ 25 Earth years
   ○ 14 Earth years
   ○　1 Earth year

6. The last planet in the solar system is now _____.

   ○ Jupiter
   ○ Mars
   ○ Neptune
   ○ Uranus

Vocabulary

A. Choose the correct word to complete each sentence.

Word Box				
exploration	cycle	flare	satellite	ripples
biome	spews	climate	atmosphere	

1. At night you can sometimes see a _____ moving across the sky.

2. The layer of air that surrounds Earth is called the _____.

3. The four seasons repeat themselves in a yearly _____.

4. Lynn threw a pebble in the pond and watched the little _____ form.

5. The _____ in Alaska is cold and snowy.

6. Jim sent up a _____ in hopes that the rescuers would see it.

7. The volcano _____ gas and ash from time to time.

8. _____ is a good way to learn about new places.

9. Many colorful birds live in the tropical rainforest _____.

B. Can you name the planets in order? It's easy if you can say a sentence in which each word begins with the same letter as a planet. Here's an example:

Many **v**ery **e**xcited **m**onkeys **j**umped **s**uddenly **u**pon **N**ancy's **p**urse.

Make up your own sentence to help you remember the planets in order. Write your sentence here.

_____

_____

Name _____

A Planet Report

Choose one planet that you would like to learn more about. Read about the planet in an encyclopedia or other reference book. Write a paragraph telling what the planet is like. Also tell whether you would like to visit this planet and the reasons why or why not.

Draw a picture of the planet.

_____

_____

_____

_____

_____

_____

_____

_____

_____

_____

_____

_____

_____

When the Dragon Swallows the Sun

The day is bright and clear. Sunshine pours over the land. People are working in the fields. The birds are chirping in the trees. All is peaceful and as it is supposed to be. Suddenly, there is a change. The light grows strange and dim, even though it is midday. The birds fall silent in the trees. The people look at the sky. A dark shape is moving across the sun's face. The day grows darker and darker. It almost seems like night.

The people are terrified. What is happening? Why is the sun growing dark? They run to their homes. They gather inside and talk with great excitement. Has something swallowed the sun? Some say it must be a dragon. Only a dragon is large enough to swallow the sun! Others argue that it cannot be so.

As they argue, the day grows bright again. The people laugh with relief. Everyone returns to work. But still, their minds are full of wonder. What happened? Did a dragon really try to swallow the sun?

©2002 by Evan-Moor Corp.

Read and Understand, Science • Grades 3–4 • EMC 3304

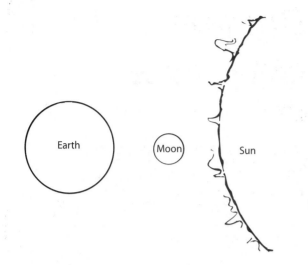

Earth   Moon   Sun

Sometimes, ancient people made up stories to explain the events of nature. This story shows one way that people long ago explained a solar eclipse. A solar eclipse takes place when the Moon passes between the Sun and the Earth. For a few minutes, the Moon blocks out much of the Sun's light. The Moon makes a large shadow on the surface of the Earth. It is this shadow that causes the day to grow dim. As the Moon continues on its orbit, the shadow moves away because the Earth and Moon keep moving. An eclipse lasts only a few minutes.

Some ancient people did study astronomy. Astronomy is the study of celestial bodies. More than 2,000 years ago, the Chinese knew the real cause of eclipses. Clay tablets have been found in Babylonia. They are over 3,000 years old. They show that the Babylonian people kept records of eclipses. Egyptians understood a lot about the movements of the objects in the sky as well. Tomb paintings show pictures of the stars and planets. A Greek astronomer named Ptolemy lived more than 2,000 years ago. We still have the books that he wrote. These books show that he understood eclipses. He was even able to predict when they might occur. We know that people have been studying the sky for a long time.

Today, scientists still study eclipses. They want to learn more about the movements of the Sun, Moon, and planets. They can tell exactly when a solar eclipse will occur. Sometimes an eclipse is total. This is when the Moon covers the Sun completely. At other times the Moon may cover only part of the Sun. This is called a partial eclipse. Scientists also know what location on Earth will be best for viewing an eclipse. Sometimes people travel long distances to experience a total eclipse of the Sun.

People now know that there is nothing to fear from a solar eclipse. We know that there are no monsters roaming the sky. No dragons try to swallow the Sun. But a solar eclipse is still very exciting. When a dark shape covers the Sun, and the day grows strangely dim, people still feel a sense of wonder.

Name _____

Questions about
When the Dragon Swallows the Sun

1. Why do you think ancient people made up stories to explain the events of nature?

   _____

   _____

2. What really happens during a solar eclipse?

   _____

   _____

   _____

3. How do we know that people of long ago studied the movements of the stars and planets?

   _____

   _____

   _____

4. What is a partial solar eclipse?

   _____

   _____

5. Why do you think some people are willing to travel a long way to experience a solar eclipse?

   _____

   _____

Vocabulary

A. The words in the box can be matched to make pairs of synonyms. Fill in the blanks using the words from the box.

Word Box					
amazement	peaceful	dim	grave	see	dark
wonder	total	frightened	disagree	complete	tomb
terrified	old	calm	ancient	argue	view

_____ means about the same as _____

_____ means about the same as _____

_____ means about the same as _____

_____ means about the same as _____

_____ means about the same as _____

_____ means about the same as _____

_____ means about the same as _____

_____ means about the same as _____

_____ means about the same as _____

B. Four ancient civilizations are mentioned in the story. Name them. You may have to change some of the words. For example, "American" would be changed to "America."

_____   _____

_____   _____

Name _____

Write Your Own Story

Make up a story that explains some event in nature in an imaginative way.

_____

_____

_____

_____

_____

_____

_____

_____

_____

_____

_____

_____

_____

_____

_____

Hailstorms and Hailstones

Imagine that you are driving along a country road in Kansas. Suddenly, it seems that thousands of baseballs are falling from the sky. They slam into your car. Bam! Bam-bam-bam! They bounce off the pavement in front of you. Crack! The windshield shatters.

Does this sound like a scary dream or maybe a bad movie? Actually, it's real life. It's called a hailstorm.

Hail is made of frozen rain. But how do raindrops get to be the size of golf balls, chicken eggs, or even baseballs? Hail forms when strong winds blow raindrops back up into the clouds. These winds are called updrafts. They carry the raindrops high up in the clouds where the air is very cold. The raindrops freeze. They pick up bits of ice and snow. They become heavier, and start to fall again.

As these frozen drops fall, they may be caught by another updraft. Again they go high in the cloud where they pick up more layers of snow and ice. They grow bigger and heavier, and begin to

fall again. If the updrafts are strong, this can happen over and over again. The drops of rain become balls of ice and snow. These balls are called hailstones. With each trip up into the clouds, the balls grow bigger and bigger. Finally, they are too heavy for the winds to lift and they fall to Earth.

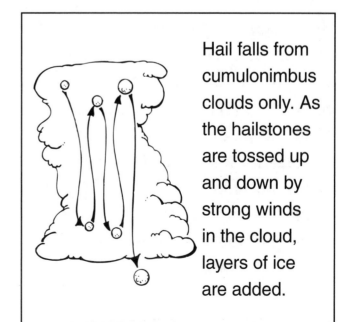

Hail falls from cumulonimbus clouds only. As the hailstones are tossed up and down by strong winds in the cloud, layers of ice are added.

When hailstones fall, they sometimes do terrible damage. A farmer's crops may be completely destroyed by hail. In minutes, tall corn can be knocked flat. Waving wheat is pounded into the soil. And just imagine what might happen to a field full of tomatoes!

Hail can strip the leaves and fruit from trees, and can even break off limbs. Large hailstones can dent metal and break glass. They can shatter windows in cars and houses.

Large hailstones can injure or kill animals that have no shelter. Hailstones have killed some people, too.

Of course, hail makes the roadways very dangerous. It can pile up on the streets and make drivers lose control of their cars. Imagine driving on a bed of slippery marbles!

Hail occurs during thunderstorms and tornadoes. It falls most often in the spring. In the United States, hail is most common in Wyoming and Colorado. It often falls on the plains of Kansas, Nebraska, Oklahoma, and Texas. Hail almost never falls over the oceans. It doesn't seem to fall in the areas of the North and South Poles. It also is very rare in hot, tropical areas.

Have you ever seen a hailstorm? When you are safe and warm inside your house it can be fun to watch the icy balls pelting the ground.

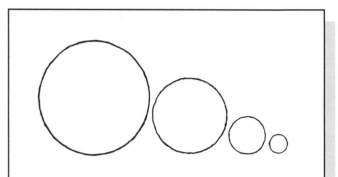

Hailstones vary in size and shape. The smallest hailstones are the size of peas. The largest hailstone on record fell in Coffeyville, Kansas, in 1970 and was nearly 18 centimeters in diameter. That's almost as big as a soccer ball!

Not all hailstones are ball-shaped. Some are flattened disks. Some are like lumpy potatoes. Some are nearly cone-shaped. If you cut a large hailstone in half, you would see the layers of snow and ice. The layers look somewhat like the rings of an onion.

Name _____

Questions about
Hailstorms and Hailstones

1. Hail is made of _____.

   ○ rocks

   ○ frozen rain

   ○ golf balls

   ○ baseballs

2. Hailstones can _____.

   ○ ruin crops

   ○ damage cars

   ○ injure animals

   ○ all of the above

3. Hail usually occurs during _____.

   ○ sunny days

   ○ snowstorms

   ○ thunderstorms

   ○ tropical rain showers

4. The largest hailstone on record was almost as big as _____.

   ○ a softball

   ○ a beach ball

   ○ a tennis ball

   ○ a soccer ball

5. Hail forms when strong winds blow raindrops high into the clouds.
   These winds are called _____.

   ○ updrafts

   ○ downdrafts

   ○ cloud drafts

   ○ hail-makers

Name _____

Vocabulary

Write the correct word in each sentence.

Word Box				
damage	flattened	shatter	crops	dent
layers	destroyed	injure	slippery	pelted

1. The tornado did some _____ to our house.

2. A shopping cart rolled across the parking lot and hit my car, leaving a large

   _____.

3. Seymour heard the window _____ just moments after he
   threw the ball.

4. Uncle Heywood planted wheat, soybeans, and some other

   _____ on his farm.

5. The fans _____ the singer with flowers after her fabulous
   performance.

6. The coat of polish made the floor very _____.

7. Kelly cried when she saw that the dog had _____ her
   homework.

8. I did not want to _____ myself, so I climbed the ladder very
   carefully.

9. The heavy boards _____ the box underneath them.

10. My chocolate birthday cake had four _____ with frosting
    between each.

Name _____

Hailstorms
Hailstones

Many Different Sizes

Hailstones are not really stones. They are balls of frozen rain. They come in many different sizes. Measure the diameter (the distance across the widest part) of each hailstone pictured.

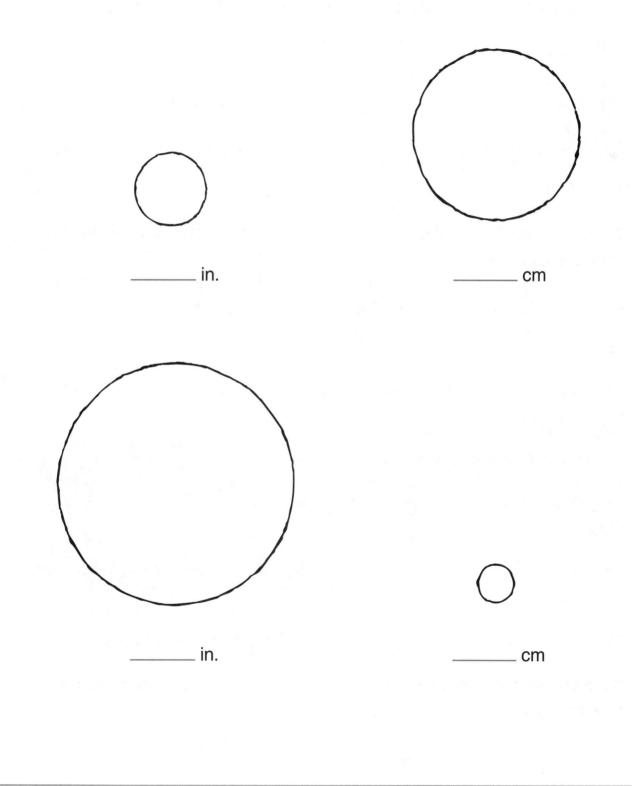

_____ in.

_____ cm

_____ in.

_____ cm

The Miracle of Light

Light is wonderful and amazing. Light from the Sun warms the Earth. It provides the **energy** plants need to grow. Plants make food for animals. Plants and animals provide food for people and other animals. Without the Sun's light, there would be no life on Earth. Light from the Sun is the most important light of all.

But light also comes from other sources. At night you can see stars twinkling in the sky. Like our Sun, stars give off their own light. (In fact, our Sun is a star!) Stars seem small and dim because they are so far away. The Moon appears to give off light, but it does not make its own light. It can only reflect light from the Sun.

Other lights also appear in the sky. Lightning zigzags through storm clouds, making a very exciting light. People who live near the North and South Poles sometimes see glimmering bands or streaks of light in the sky. In the **Northern Hemisphere** this is called the aurora borealis, or "northern lights." In the Southern Hemisphere it's called the aurora australis or "southern lights."

Nature gives us another surprising source of light. Some animals are able to give off light from their bodies! Have you ever seen fireflies dancing in the air on a summer evening? Chemicals inside the fireflies' bodies react to give off light. This makes it look like little lights on their abdomens are blinking on and off. Squid and many other sea animals can also give off light.

These lights from nature add beauty and wonder to our lives. But many of the lights we see every day do not come from nature. These are manmade, or artificial, lights. Light bulbs, laser beams, and flashlights are all examples of artificial lights.

Artificial light is important. It allows people to work and travel at night. Many stores and factories stay open all night long. We can go shopping at night, or to the library. Headlights on our cars allow us to see the road at night. Streetlights help us to walk safely along the sidewalk. Airplanes take off and land on lighted runways.

Artificial lights make our homes safer and more comfortable. They make our lives easier and more fun. In the past, people had only candles and firelight. They often went to bed when the sun went down. Today we only have to flip a switch to get light. We can read, work, and play for many more hours each day. Manmade lights even give us entertainment such as movies, video games, and TV.

Artificial light has another important job. It can be used to carry information from one place to another through special cables. Many telephones and computers are connected to these cables. Light can help us talk to people who are far away. It can help us use the Internet.

Daytime sky   **Nighttime sky**

Take a moment to notice light. Go outside on a sunny day. Feel the warmth on your skin. Gaze at the brilliant stars in the night sky. Indoors, count the number of artificial light sources that you and your family use. Open the door to your refrigerator and see everything that's inside. Enjoy the miracle that is light!

Name _____

Questions about
The Miracle of Light

1. What is the most important source of light?

   ○ a light bulb

   ○ a flashlight

   ○ car headlights

   ○ the Sun

2. The northern lights and southern lights look like _____.

   ○ lightning bolts

   ○ streaks of colorful light

   ○ moonlight

   ○ fireflies

3. Fireflies and some other living creatures make lights using _____.

   ○ chemicals in their bodies

   ○ friction of their wings

   ○ sounds

   ○ magic

4. Artificial lights are manmade. Which of the following is **not** an artificial light?

   ○ light from the television

   ○ laser beam

   ○ starlight

   ○ light from a streetlight

5. The stars we see at night seem small because _____.

   ○ they are small

   ○ they are reflecting light from the Sun

   ○ they are twinkling

   ○ they are so far away

Name _____

What Is It?

Label each picture with a word from the box.

Word Box			
zigzag	northern hemisphere	aurora borealis	south pole
squid	lightning	abdomen	firefly

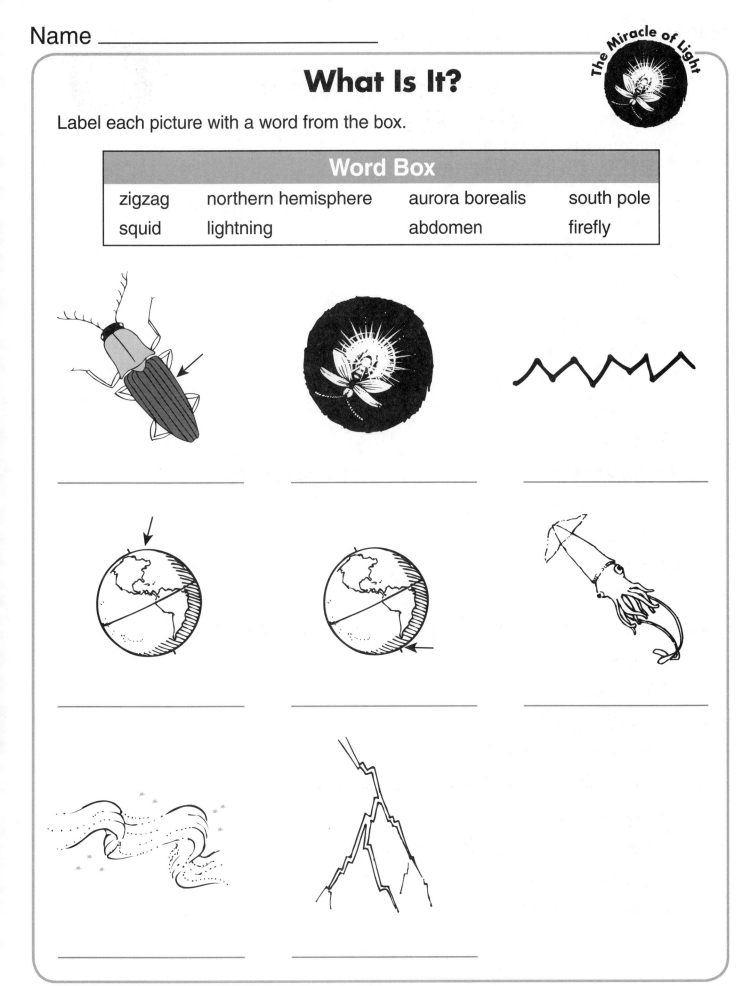

Name _____

A List of Lights

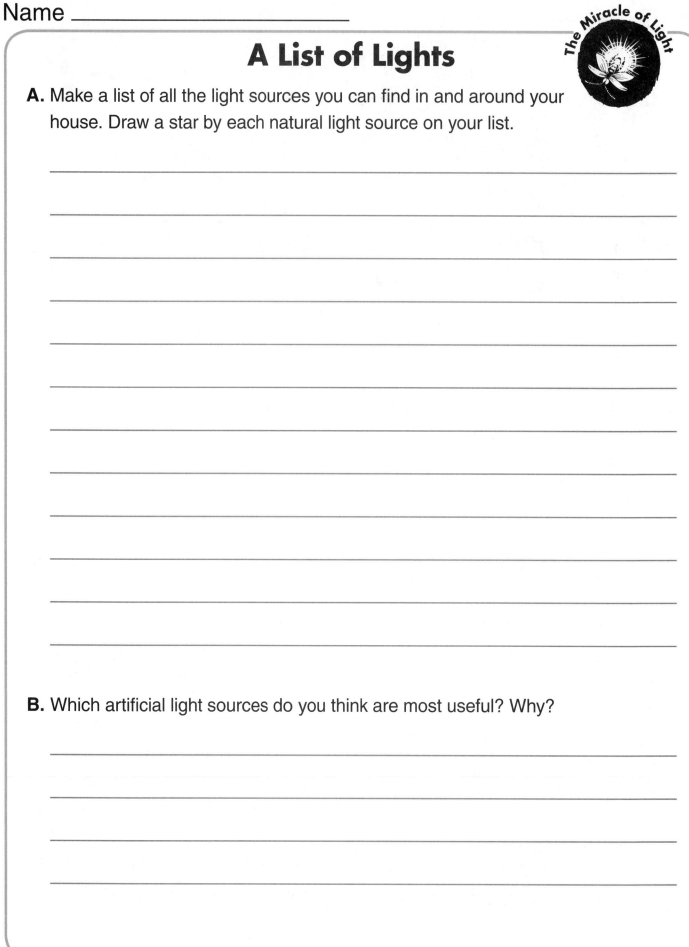

A. Make a list of all the light sources you can find in and around your house. Draw a star by each natural light source on your list.

_____

_____

_____

_____

_____

_____

_____

_____

_____

_____

B. Which artificial light sources do you think are most useful? Why?

_____

_____

_____

_____

A Class by Itself

wolf

German shepherd

Can you tell the difference between a wolf and a German shepherd? These animals look a lot alike. Each has four legs and a tail. Each has fur. And females from each **species** nurse their young with milk. But a wolf and a German shepherd have different features. These features help scientists tell one group of animals from another.

The division, or **classification**, of animals and plants into groups is called **taxonomy**. Scientists classify animals by looking at things they have in common.

The first scientist to classify animals lived thousands of years ago. His name was **Aristotle**. Aristotle lived in Greece from 384 to 322 B.C.E. Aristotle saw that animals could be classified by common traits. He identified four ways to group animals. These were by way of living, actions, habits, and body parts.

Aristotle began dividing animals into groups. Some of these groups were birds, whales, fish, and insects. Then Aristotle broke each large group into smaller groups. He wrote that animals with two feet were different from animals with four feet. Some animals had hair or feathers. Some did not. Animals with shells were different from animals without shells. These classifications let Aristotle identify different groups of animals.

Another important scientist was Carolus **Linnaeus**. Linnaeus was born in Sweden in 1707. At that time many new animals were being discovered. **Biologists**, scientists who study living things, had a hard time placing these new animals into Aristotle's system. Some animals didn't fit well into any of Aristotle's groups. So Linnaeus made up a new system. Linnaeus's basic system of classification is still used today.

Scientists look at many things when they are classifying animals. The easiest way to classify an animal is to look at its body. It's easy to see that a cheetah and a leopard look a lot alike. So these animals are grouped together. Other times, scientists have to look very closely at the animals' bodies to see what things are alike. That's why some animals that don't look alike can be part of the same group.

Scientists also look at where animals live and what they eat. A bird that eats insects is different from a bird that eats nuts or seeds. Polar bears and sun bears are both bears. But polar bears live where it is very cold. Sun bears live where it is hot. Although both animals are bears, they are classified in different groups.

Classifying animals is like fitting pieces into a puzzle. Each animal fits into a special place. It's up to biologists to find out just where that place is.

Gray Squirrel

This is how a gray squirrel would be classified in the Linnaean system.

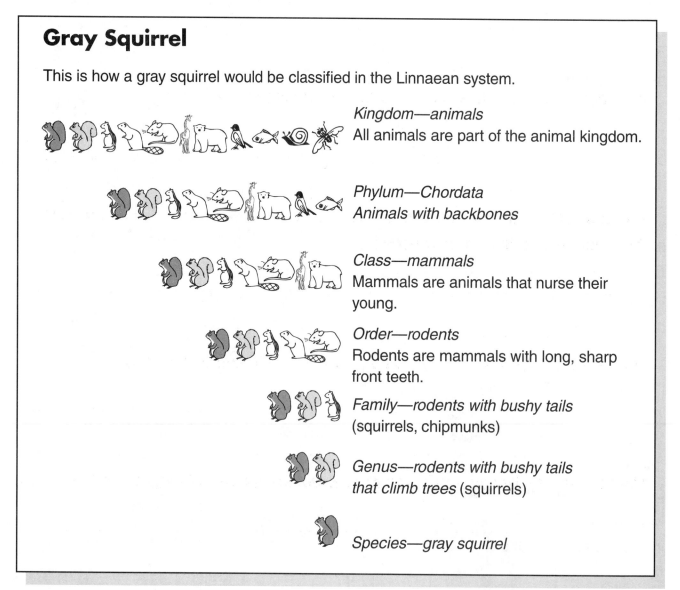

Kingdom—animals
All animals are part of the animal kingdom.

Phylum—Chordata
Animals with backbones

Class—mammals
Mammals are animals that nurse their young.

Order—rodents
Rodents are mammals with long, sharp front teeth.

Family—rodents with bushy tails
(squirrels, chipmunks)

Genus—rodents with bushy tails
that climb trees (squirrels)

Species—gray squirrel

Questions about *A Class by Itself*

1. How do scientists classify animals?

_____

_____

2. Why was Aristotle mentioned in this story?

_____

_____

3. What were the four things that Aristotle used to group animals?

_____

_____

4. Why did Linnaeus make up a new system to classify living things?

_____

_____

5. When scientists are deciding what group an animal belongs to, what things do they look at besides appearance?

_____

_____

6. Name two things that are the same about squirrels and chipmunks.

_____

_____

7. Name one way that squirrels and chipmunks are different from each other.

_____

_____

Name _____

Vocabulary

Use words in the box to complete the crossword puzzle.

Word Box					
Aristotle	Linnaeus	traits	system	mammals	classification
species	biologist	squirrel	rodents	taxonomy	identify

Across

4. characteristics
5. dividing animals and plants into groups
7. a specific type of animal or plant
9. to tell what something is
10. a way of organizing things
11. animals with long, sharp front teeth
12. the process of dividing into groups

Down

1. the first scientist to classify animals
2. a scientist who studies animals
3. animals that nurse their young
6. created the system of plant and animal classification we use today
8. a rodent with a bushy tail that climbs trees

Name _____

Classify Your Classroom

Get together with a group of three classmates and make up a system for classifying objects in your classroom. Begin with general categories, such as "manmade," "natural," "consumable," or "nonconsumable." Continue making the groups more specific until you end up with a single object. (For example: consumable—paper—construction paper—red construction paper.)

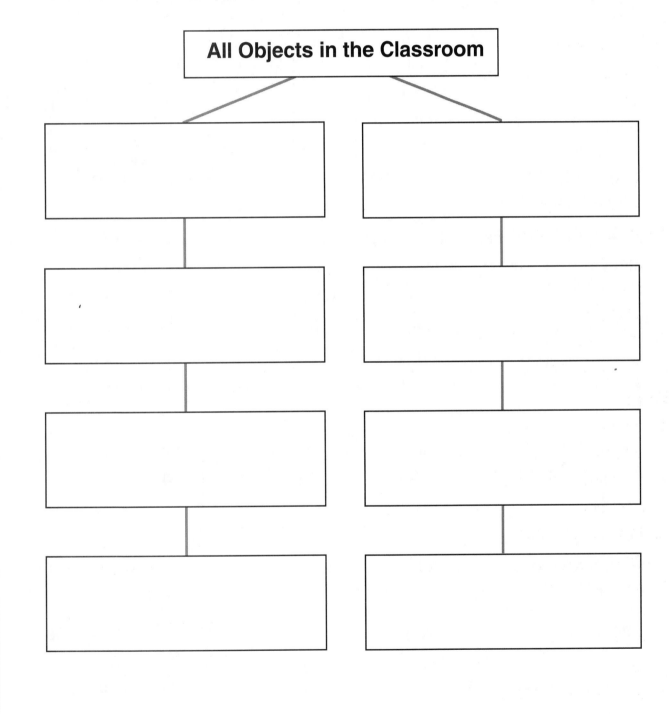

All Objects in the Classroom

At Home in the City

Where do coyotes and raccoons live? If you said the forest or the prairie, you would be right! Did you know that these animals also live in the city? Many animals that live in the wilderness have also learned to live in cities.

As more houses, offices, stores, and roads are built in what used to be wilderness, animals must learn to live in new places. Sometimes the animals cannot do this. Maybe they cannot find food in their new habitat. Or maybe they can't find a safe home to raise their young.

However, the city can make a good home for some animals. There are many places to live in the city. Some animals live under bridges or behind billboards. Some live in empty buildings. Others live in parks or empty lots.

For some kinds of animals, the city provides lots of food. City animals eat insects, mice, and birds. They also steal food from garbage cans. Some animals snack on people's gardens. City animals drink water from lakes in parks.

Some wild animals live in cities because there are few predators there. A raccoon might be prey for a wolf in the wild. In the city, there are no wolves.

Many raccoons live in the city. These animals are very smart. They can learn how to open doors and unscrew jars. Raccoons can also take the lids off garbage cans. These skills help the raccoon find food in the city.

Raccoons eat almost anything. Raccoons like fruit, insects, birds, mice, and fish. They will eat vegetables and fruit they find in garbage cans. They will even eat cookies and ice cream! Eating lots of different foods allows raccoons to survive in the city.

There are lots of places for raccoons to live in the city. Their sharp, curved claws help them climb. City parks are full of trees. These trees make good homes for raccoons.

Raccoons also live in other places. Some raccoons make nests in the chimneys of houses. Others climb fire escapes to the top of tall buildings. Then they move into the heating vents on the roof. Still other raccoons live in garages, sheds, and empty buildings. A city raccoon can find many places to call home!

Another animal that sometimes lives in the city is the coyote. It's easy for this animal to adapt to a new habitat. The coyote lives in more places today than it did 100 years ago.

Coyotes have no trouble finding food in the city. That's because a coyote will eat almost anything. Coyotes dig up gardens. They look through garbage cans for leftover vegetables and meat. They steal fruit from boxes behind grocery stores.

Coyotes are also good hunters. A coyote will eat most of the other animals that live in the city. Squirrels, birds, raccoons, and insects are all part of the coyote's diet. A coyote will even eat cats and small dogs. A coyote that lives in the city is often bigger than a coyote that lives in the country. That's because a city coyote can find so much to eat.

Parks make good homes for city coyotes. Other coyotes live in empty lots or alleys. People who live in cities might pass a coyote's home every day and never know it!

Of course, most wild animals don't belong in the city. Wild animals can hurt people or pets if they become frightened. People get angry when wild animals dig up their gardens, knock over their garbage cans, and hurt their pets. Wild animals can also spread diseases, such as rabies. The best place for wild animals to live is still in the wild!

Name _____

Questions about
At Home in the City

1. Which of these animals does **not** live in the city?

   ○ raccoon
   ○ wolf
   ○ squirrel
   ○ coyote

2. Where can city animals find food?

   ○ parks
   ○ garbage cans
   ○ gardens
   ○ all of the above

3. Which of the following is **not** food for a raccoon?

   ○ insects
   ○ fruit
   ○ tree bark
   ○ cookies

4. Coyotes can live in the city because _____.

   ○ they like people
   ○ there is plenty of food
   ○ it is crowded
   ○ they always live in parks

5. Coyotes live in _____.

   ○ alleys and parks
   ○ the rainforest
   ○ houses
   ○ apartments

6. More animals will live in cities today because _____.

   ○ people want them to
   ○ cities are taking up more land
   ○ it is a better place for them than the wild
   ○ there is no food in the wild

Name _____

Vocabulary

A. Draw a line between the word on the left and its meaning on
the right.

wilderness   an animal that hunts other animals

adapt   a place where an animal lives

predator   a place where no people live

habitat   an animal that is hunted by other animals

prey   to change because you are in a new situation

B. Use the vocabulary words above to complete these sentences.

1. Mice are often _____ for larger animals.

2. Some wild animals can _____ to life in the city.

3. A wolf is a fierce _____.

4. The city is a good _____ for raccoons.

5. The _____ is not as crowded as the city.

C. Draw a picture of a wild animal that has adapted to life in the city. Label your
drawing.

Name _____

Graphing Information

Some scientists study animals that live in the city. A scientist watched a coyote for one week. Then he made a graph of all the food the coyote ate. Look at the graph and answer the questions.

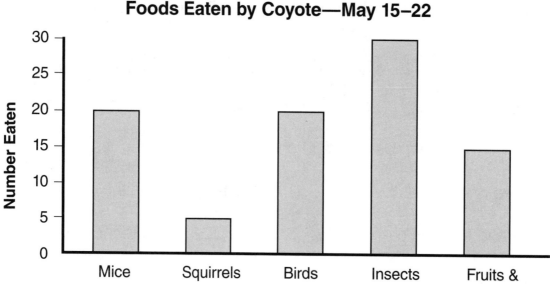

Foods Eaten by Coyote—May 15–22

1. How many different categories of food are shown on this graph? _____

2. How many mice did the coyote eat this week? _____

3. How many squirrels did the coyote eat this week? _____

4. Did the coyote eat more birds or more insects? _____

5. What else can you tell from this graph?

_____

_____

Making Old Things New

Stop! Don't throw that empty milk jug in the trash! Like many other kinds of plastic, that jug can be **recycled** into something new.

Recycling is the process of collecting used items and using the materials they're made out of to make new items. Many different items can be recycled, including glass, plastic, aluminum, newspapers, magazines, paper, motor oil, and cardboard. Items that can be recycled are called **recyclables**.

Every year tons of garbage is dumped in **landfills**. Then dirt is piled on top of the garbage. People once thought that the buried garbage would **decompose**, or break down, under the ground. But decomposition can't happen without air and moisture. The garbage can't decompose underground because there isn't enough air or moisture there. Instead, the garbage just sits there.

Another problem with landfills is that they take up a lot of space. The garbage keeps on coming, but there are fewer places to put it. One reason communities have started recycling is to help ease the garbage problem.

Recycling became popular in the United States during World War II. Communities had "scrap drives" to collect tin cans and old metal. These items were used to make machines used in the war.

When the war ended in 1945, most people stopped recycling. Then, in the early 1970s, scientific studies showed that **pollution** was damaging the **environment**. Many people felt that recycling was something they could do to help reduce pollution. Today, many communities ask their residents to recycle.

One of the most common recyclables is plastic. Everything from soda bottles to plastic bags can be recycled.

The first step in the recycling process is to take the recyclable out of the **waste stream**. Instead of throwing out recyclables, people place them in recycling bins. These bins are usually given out by the town. Sometimes they come from the community's garbage or **recycling contractor**. On certain days the bins are placed outside. Then trucks take the recyclables to a recycling center. Other communities ask people to take their recyclables to the recycling center.

At the recycling center, the recyclables are washed to get rid of any dirt. Then the plastics are placed on large conveyor belts for sorting.

Have you ever seen the number inside the recycling symbol on a soda bottle or plastic bag? That number tells you what type of plastic the item is made of. In terms of recycling, there are seven different types of plastic. At the processing center, each type of plastic is treated a different way.

After the plastic recyclables are sorted, trucks take them to processing centers. Some plastics are melted down in huge furnaces. Then the liquid plastic is molded into new items. Melting and remolding is the best process for producing thick, heavy plastics. Detergent bottles are a good example of a plastic that can be melted down and remolded.

Other plastics are shredded or ground into tiny pieces. These pieces can be used as insulation. Some recycled plastic is used to fill sleeping bags or jackets.

Recycling plastic and other materials is not as easy or cheap as burying them in a landfill. But recycling provides a cleaner way for us to get rid of garbage. Why throw something away when you can use it again?

Name _____

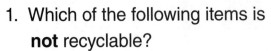

Questions about
Making Old Things New

1. Which of the following items is **not** recyclable?

   ○ plastic
   ○ newspaper
   ○ glass
   ○ food

2. Garbage buried in landfills can't break down because _____.

   ○ it's too dirty
   ○ there isn't enough air or moisture
   ○ there isn't enough light
   ○ it's too wet

3. During the 1970s, people began to worry about damage to the environment from _____.

   ○ pollution
   ○ recycling
   ○ animals
   ○ weather conditions

4. Dividing recyclables into different types is called _____.

   ○ contracting
   ○ sorting
   ○ washing
   ○ molding

5. Which of the following is a way to recycle plastic?

   ○ shredding
   ○ melting
   ○ grinding
   ○ all of the above

6. Recycling is better than throwing garbage away because _____.

   ○ it is cheaper
   ○ it is easier
   ○ it is better for the environment
   ○ it is fun

Name _____

Vocabulary

Write the word from the box that matches each definition.

Word Box			
recycling	waste stream	landfill	contractor
decompose	pollution	environment	recyclables

1. to break down or rot __ __ __ __ __ __ __ __ __
                               2   9

2. someone hired to do a job __ __ __ __ __ __ __ __ __ __
                                   7   8

3. items that can be made into new items

   __ __ __ __ __ __ __ __ __ __ __
       14   11

4. damage to water, air, and soil __ __ __ __ __ __ __ __ __
                                      13   5

5. the world of nature; land, sea, and air

   __ __ __ __ __ __ __ __ __ __ __
       10

6. collecting used items and changing them into something new

   __ __ __ __ __ __ __ __ __
       1   4

7. items that are thrown away become part of this

   __ __ __ __ __ __ __ __ __ __ __
       12   3

8. a place where garbage is buried __ __ __ __ __ __ __ __
                                   6

Match each numbered letter above with the same number in the puzzle below to discover an important fact.

__ __ __ __ __ __ __ __ __   __ __ __ __ __ __ __
10 3 1 14 1 6 5 7 4   10 3 2 13 1 3 11

__ __ __ __ __ __ __ __ __
9 8 6 6 13 12 5 8 7

©2002 by Evan-Moor Corp.   Read and Understand, Science • Grades 3–4 • EMC 3304

Sequencing Events

Making Old Things New

A. Number the following events in the order in which they occur during the recycling process.

_____ Recyclables are washed.

_____ Plastic recyclables are melted down or shredded into new items.

_____ People put recyclables in recycling bins.

_____ Recyclables are sorted.

_____ The recycling contractor picks up recyclables.

_____ Recyclables are taken to a processing center.

_____ Trucks take the recyclables to the recycling center.

B. Draw a picture to show how you participate in recycling.

Mountains

Folded Mountain

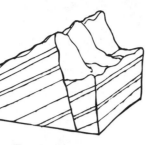

Fault-block Mountain

Dome Mountain

Volcanic Mountain

Mountains are formed in several different ways. But all mountains are made by movements and changes in the Earth's **crust**. The Earth's crust is the outer skin, or shell, of the Earth. This crust is not one solid piece. Instead, it is made of several large pieces, or plates.

These plates are always moving, but so slowly that we cannot tell. The movements of the plates cause sections of the crust to bend or break. Where the crust bends or breaks, mountains are sometimes formed. This can happen in four different ways.

Sometimes areas of the crust buckle or fold. Mountains formed in this way are called **folded mountains**.

In other cases, pressure from movement in the crust causes huge cracks to form. These cracks are called **faults**. There is a lot of pressure along a fault. The crust can suddenly break into huge blocks. Some of the giant blocks

jut upward, while others slip downward. Mountains formed in this way are called **fault-block mountains**.

Dome mountains are formed when the crust simply bulges upward. The bulging is caused by pockets of molten rock that push upward toward the surface, causing the crust to rise. An example of this kind of mountain can be seen in the Black Hills region of South Dakota.

Volcanic mountains are formed when **magma**, the molten rock below the Earth's crust, breaks through to the surface. Once the molten rock, or magma, reaches the Earth's surface, it is called **lava**. Lava is molten rock above the Earth's crust. The molten rock pours out, and gas and ash explode from the **vent**, or opening. Layers of cooled ash and lava form a large cone around the vent. These layers pile up, one on top of the other, building the mountain higher and higher. The mountains of the Hawaiian Islands were formed by volcanoes.

As soon as a mountain forms, it begins to wear away. This process is called **erosion**. Rain falls on the mountain. The water flows down the mountain in streams that carry away tiny bits of rock. Snow falls on the top of the mountain. If the mountain is high enough, some of the snow will remain there all year. Over time the snow hardens into ice. Year after year, the snow gets deeper and heavier. After many years, the layers of ice form into **glaciers** that move slowly down the mountain, carving out bowls and valleys. The glaciers pick up rocks that grind away at the mountainside like sandpaper.

Nature works on the mountain in other ways, too. Trees send roots into little cracks, or crevices, in the rock. Water seeps into these crevices, too. In cold weather the water freezes and expands, enlarging the cracks. The roots grow bigger and exert more pressure on the rock. The action of roots and ice causes pieces of the rock to break away. Even the wind can scrub at the mountainside, slowly wearing it away.

Because of this erosion, older mountains appear smoother, lower, and more rounded than younger mountains. Older mountains are often covered with trees and plants. Younger mountains are sharper, more rugged, and taller. They are often rocky and covered with snow.

Looking up at one of these towering giants, it is hard to imagine that wind, rain, and snowflakes can ever make it disappear! It may take millions of years, but in time the mountain will wear away completely. At the same time, new mountains will be forming on the ever-changing surface of our planet.

Younger mountains are sharper and taller.

Older mountains are smoother and lower.

Name _____

Questions about Mountains

1. The Earth's crust _____.

   ○ is made of one solid piece

   ○ is made of several large pieces

   ○ does not move

   ○ is very hot

2. Mountains formed when ash and molten rock pour out of the Earth are called _____.

   ○ fault-block mountains

   ○ folded mountains

   ○ dome mountains

   ○ volcanic mountains

3. Erosion can be caused by _____.

   ○ wind

   ○ water

   ○ ice

   ○ all of the above

4. Fault-block mountains are formed when _____.

   ○ large chunks of crust break and move

   ○ volcanoes erupt

   ○ the crust folds and bends

   ○ snow collects and forms glaciers

5. Younger mountains are usually _____.

   ○ smoother and rounder

   ○ smaller

   ○ sharper and more rugged

   ○ covered with trees

6. Which of these mountain ranges is an example of dome mountains?

   ○ Rocky Mountains

   ○ Black Hills of South Dakota

   ○ Appalachian Mountains

   ○ Sierra Nevada

Mountains

Vocabulary

A. Write each word on the line in front of its meaning.

Word Box				
lava	crust	dome	glacier	vent
faults	crevice	cone	plates	magma

1. _____ a small crack

2. _____ the outer layer of the Earth

3. _____ the opening of a volcano

4. _____ the large pieces of the Earth's crust

5. _____ molten rock below the Earth's crust

6. _____ a shape with a round base and a pointed end

7. _____ a rounded shape, somewhat like half of a ball

8. _____ a large mass of snow and ice

9. _____ huge cracks in the Earth's surface

10. _____ molten rock above the Earth's crust

B. What does **erosion** mean?

_____

_____

C. In your own words, explain how the process of erosion can wear away
a huge mountain.

_____

_____

_____

Name _____

How Mountains Are Made

A. Label the pictures of the four different ways in which mountains are formed.

_____   _____

_____   _____

B. Show this page to someone in your family. Explain how each kind of mountain is formed.

The Magic Eye

A patient is rushed to the emergency room of a large hospital. He says his stomach hurts. The doctor needs to look inside the man's body to find out what's wrong. How can he do that? He can use a special machine called a CT scanner.

CT is short for **computerized tomography**. A CT scan is a special kind of X-ray image. It shows doctors what's deep inside a patient's body.

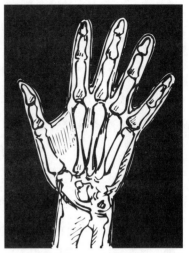
X-ray of a hand

Traditional X-ray images were first used in 1896. X-ray images are formed by passing **radiation** through the body. The rays pass through soft body tissues, but not through bones. The rays then strike photographic film, creating a picture of the bones. That picture can show a doctor if something is wrong with the bones.

X-rays work well on **dense** parts of the body, like bones. For example, an X-ray can show a **fracture** in an arm or leg bone. But X-rays are not very good at showing other parts of the body. Soft tissues, such as blood vessels and muscles, don't show up well on X-rays. Neither do **organs**, such as the brain or heart.

During the 1960s, two scientists worked to make X-rays better. One was Sir Godfrey Newbold Hounsfield. The other was Allan Macleod Cormack. These two men didn't know each other, but they each came up with the idea for the CT scanner.

Both Cormack and Hounsfield thought of sending many X-rays through the body at different angles. They thought this would let doctors take pictures of body organs. Then a computer could combine the images into one picture. In 1972, a British company used this idea to create the first CT scanner.

©2002 by Evan-Moor Corp.
Read and Understand, Science • Grades 3–4 • EMC 3304

Doctors began using CT scans to help patients. In the past, if a patient had something wrong deep inside the body, doctors had to perform an operation in order to find the problem. Today, doctors use CT scans to **diagnose** many different medical problems. Operations can be dangerous and painful. CT scans are safe and painless. CT scans also allow doctors to find problems more quickly than operations do.

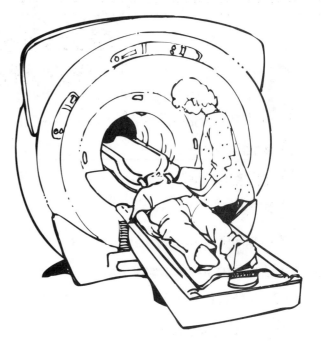

A CT scanner is about 8 feet (2.4 m) tall. The machine is shaped like a huge donut. During a CT scan, the patient lies on a special table. This table slides into the scanner. Then a tube beams X-rays through the patient's body. The scanner spins around as it sends out the X-rays. Special machines look at the X-rays and send the images to a computer. In a few seconds, the computer screen shows an image of the inside of the body.

CT scanners were first used to look at the brain. This is still the most common use of CT scans. A CT scan can show doctors if a patient has suffered a brain injury. It does this by showing blood or damaged tissue inside the brain. Doctors may take several CT scans over a few days or weeks. These scans tell them if a patient's injury is getting better or worse.

CT scans are also used to diagnose tumors, infections, and injuries deep inside the body. They can also give doctors a closeup look at body tissues during operations. This helps doctors be more precise as they work.

X-ray images have come a long way since their first use in 1896. Today, X-ray technology lets doctors do what once seemed impossible—to see inside the human body.

Name _____

Questions about
The Magic Eye

1. A CT scan is a type of _____.
   - ○ operation
   - ○ doctor's visit
   - ○ X-ray
   - ○ hospital

2. Traditional X-ray images are _____.
   - ○ a new invention
   - ○ more than 100 years old
   - ○ about 30 years old
   - ○ thousands of years old

3. X-rays are good at showing _____.
   - ○ organs
   - ○ muscles
   - ○ bones
   - ○ blood vessels

4. CT scans were first used to see _____.
   - ○ the brain
   - ○ broken bones
   - ○ the heart
   - ○ the liver

5. CT scans are _____.
   - ○ painful
   - ○ very slow
   - ○ dangerous
   - ○ painless

6. Doctors like CT scans because the scans _____.
   - ○ are fun to use
   - ○ cost a lot of money
   - ○ are asked for by patients
   - ○ help them diagnose problems

Vocabulary

Use words from the box to fill in the crossword puzzle.

Word Box				
traditional	radiation	dense	fracture	patients
organs	diagnose	operation	scanner	

Across

1. the cutting open of the body
4. to determine what disease a patient has
7. the sending out of rays of energy
8. thick
9. a computerized tomography machine

Down

2. something in use for a long time
3. people being treated by a doctor
5. body parts, such as the brain or heart
6. to break

The Magic Eye

Invent a Machine

Pretend you are an inventor. What type of machine could you invent to make someone's job easier? What does it do? How does it work? How can people use it? Describe your invention below.

_____

_____

_____

_____

_____

_____

_____

_____

_____

_____

_____

_____

_____

_____

_____

_____

Petrified Forest National Park

The Petrified Forest National Park is located in the state of Arizona. It is in the middle of a hot, dry desert. But 200 million years ago, the land was different. This area was a warm, swampy flood plain. Tall trees, ferns, and other plants grew along the banks of streams and rivers. Dinosaurs and other reptiles lived in the forest.

Some of the trees in the forest grew old and died. Some were toppled by wind or flooding. These trees settled into the earth. Some of them were buried in sand and mud. Volcanic eruptions covered the area with ashes. The minerals in the ashes mixed with the water in the ground. The mineral-rich water soaked into the trees. Over time, the wood was replaced completely by these minerals. The trees had been "petrified," or turned into fossils. They were buried deep in the earth.

Millions of years passed. Then, about 60 million years ago, this area went through another change. The land was lifted up. The petrified trees were pushed to the surface. The pressure on the trees caused them to break into pieces. Water and wind scrubbed away the soil that surrounded them. The petrified trees were left lying on the surface. Today we can see, touch, and marvel at these amazing fossils.

Dinosaur Fossils

Fossils of some very early dinosaurs and reptiles have been found in and around the park. You can see some exhibits about these animals at the Rainbow Forest Museum. There are many plant and animal fossils on display. This museum is located at the south entrance to the park.

Placerias gigas
This reptile weighed up to 2 tons (1.8 metric tons). It probably ate roots and tough plants.

Coelophysis bauri
This small dinosaur was only 8 feet (2.5 m) long and weighed about 50 pounds (22 kg). It had sharp teeth, so it probably ate meat.

Smilosuchus gregorii
This reptile looked quite a bit like a crocodile. It lived in lakes and rivers, and ate fish and other small animals.

Points of Interest

There are many places to see fossils in the Petrified Forest National Park. There are two visitor centers. One is near the north entrance. The other is near the south entrance. A 27-mile drive connects the two. You will want to make several stops along the way.

- Take the short trail to the Long Logs. Notice the bright colors of the minerals in the fossilized trees.

- Don't miss the Crystal Forest. Notice how every detail of each tree is perfectly preserved. The texture of the bark looks like living wood, even though it is really millions of years old! Many of the logs here used to contain colorful crystals. Sadly, these have been broken off and carried away by visitors.

- Stop at the Jasper Forest Overlook. You can see some trees that still have roots attached. This means that the trees must have grown nearby.

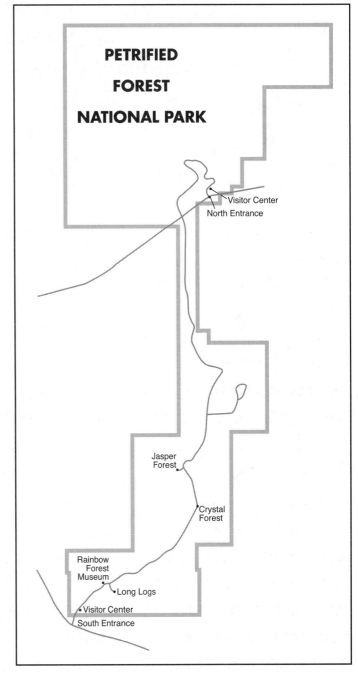

Be a Good Visitor

Please don't disturb any fossils you might find in the park. Instead, report these finds to a park ranger. Never remove pieces of petrified wood or anything else from the park. Petrified wood is for sale in many stores outside the park. Help us preserve the park so that others may enjoy it for generations to come.

For more information, check out the National Park Service Web site at www.nps.gov.

Name _____

Questions about
Petrified Forest National Park

1. Petrified Forest National Park is located in the state of _____.

   ○ Montana
   ○ Maine
   ○ Arizona
   ○ Colorado

2. This park is located in _____.

   ○ a jungle
   ○ a dry desert
   ○ a cold, snowy region
   ○ a pine forest

3. Two hundred million years ago, this area was _____.

   ○ warm and wet
   ○ cold and wet
   ○ cold and dry
   ○ warm and dry

4. The trees that grew here were turned into fossils when _____.

   ○ the wood dried out in the sun
   ○ the trees were covered with ice
   ○ mineral-rich water soaked into the trees
   ○ all of the above

5. If you find any fossils in the park, you should _____.

   ○ pick them up and take them home
   ○ pick them up and take them to the museum
   ○ leave them where they are and tell a park ranger
   ○ cover them with dirt to hide them

Name _____

Vocabulary

Circle **yes** or **no** to show whether the underlined word in the sentence has the meaning that is given. Then write a sentence of your own using the underlined word.

1. Jed waded into the <u>swamp</u> to look for snakes.

   In this sentence, **swamp** means "a marshy area of soft, wet land."   **yes   no**

   _____

2. Buried beneath layers of ash and mud, the dinosaur bones slowly became <u>petrified</u>.

   In this sentence, **petrified** means "melted."   **yes   no**

   _____

3. The <u>reptile</u> crawled up on a warm rock and went to sleep.

   In this sentence, **reptile** means "a cold-blooded animal such as a lizard or an alligator."   **yes   no**

   _____

4. The <u>ranger</u> explained the park rules to the family in the campground.

   In this sentence, **ranger** means "a person who watches over a park or forest area."   **yes   no**

   _____

5. I hope that this interesting park is <u>preserved</u> for the future.

   In this sentence, **preserved** means "destroyed."   **yes   no**

   _____

Name _____

Fossil Hunt

Fossils are clues to the past. They tell us all that we know about the plants and animals that lived on the Earth millions of years ago. Without fossils, we would not know that the strange and amazing creatures we call dinosaurs ever roamed the Earth.

Paleontologists are scientists who use fossils to study ancient plant and animal life. They search for fossils in many parts of the world, hoping to learn more from these important clues.

For each fossil picture below, write a sentence explaining what you think might have made the fossil.

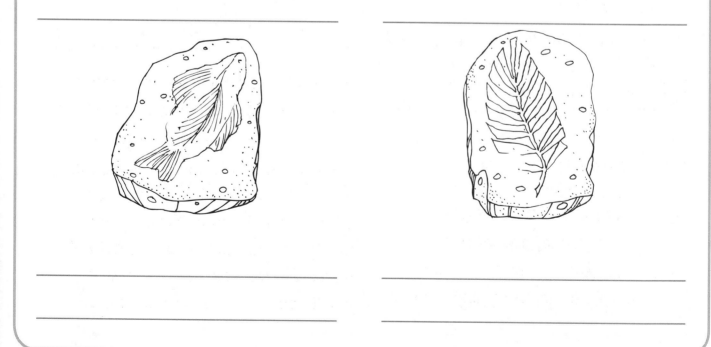

_____   _____

_____   _____

Nature's Gifts:
The Materials of the Earth

It is early morning. The first rays of the sun slip into the mouth of the cave. The family group that is sleeping inside begins to wake up. The women stir up the fire. The children gather wood. The men check their weapons, which are stout wooden spears. Each spear has a stone point at the tip. The men will hunt today. Food supplies are low. There isn't much to eat for breakfast, only scraps of meat left over from the evening meal.

The women and children will also go in search of food. They take rough baskets made of bark and reeds. While the men wander the grassland looking for game (animals that are hunted for meat), the women and children hunt for food in the forest. They gather berries in the baskets. They use sharpened sticks to dig up tasty roots. If they are lucky, there will be a feast for the whole group this night!

People have always used the materials offered by nature to make the things they need. These things make life better and easier. Even the earliest humans found ways to make useful tools. They chipped hard stone into sharp spear points. They used flexible bark to make baskets. They sharpened strong sticks to make digging tools. They used the skins of animals to make clothing and blankets.

It has been thousands of years since these early humans roamed the Earth. Our way of life has changed a great deal. Humans have learned much more about the Earth's natural materials. We have learned how to use these materials in better ways.

For instance, we have learned to take fiber from plants, like cotton, and make it into cloth. We have learned to use plants for other purposes, too. Today many products are made from plants. Oils, paint, soap, and medicines are a few of these products.

We have learned to use metals like copper, iron, gold, and silver. We use these metals to make many useful things. Metals are used to make machinery, automobiles, and airplanes. Metals are also used to make coins and jewelry. We have even learned how to mix metals to make new materials. Two or more metals mixed together form an alloy. Some alloys are very useful because they are lightweight but very strong.

Brand new materials have been developed, too. Plastics, for example, are not found in nature. They are made by mixing chemicals that are found in natural materials such as petroleum (oil), coal, and certain kinds of plants. Many different kinds of plastics can be made from these chemicals. Plastics are used to make a wide variety of objects. Ketchup bottles, trash cans, football helmets, and even parts for the space shuttles are made of plastic!

We are constantly learning more about the materials that make up our world. Every day inventors try to find new uses for nature's gifts. Every day scientists try to combine these materials in new ways. It is hard to imagine what the future holds. There will probably be many exciting discoveries in your lifetime. The materials of the Earth are truly gifts to treasure.

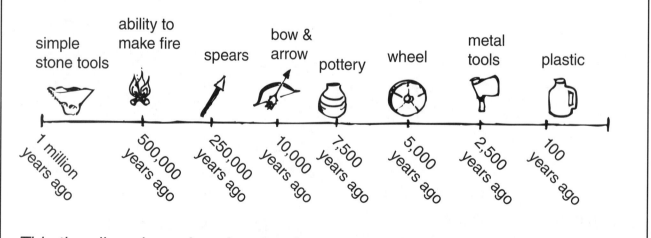

This time line shows how humans have learned to use nature's gifts over time. (Dates are approximate.)

©2002 by Evan-Moor Corp. Read and Understand, Science • Grades 3–4 • EMC 3304

Questions about
Nature's Gifts:
The Materials of the Earth

1. Next to each material, write the name of an object made by early humans.

   stone _____

   bark _____

   animal skins _____

   wood _____

2. List four products that are made today from each material.

plants	metals	plastics
_____	_____	_____
_____	_____	_____
_____	_____	_____
_____	_____	_____

3. From what are plastics made?

   _____

   _____

4. Do you think the author chose a good title for this article? Explain your thinking.

   _____

   _____

   _____

   _____

Vocabulary

We can often learn the meaning of a word by the way it is used in the sentence. This is called the "context." Use context clues to complete the exercises below.

1. The soldier cleaned his **weapon** before the battle began.

   Circle the name of the object that is **not** usually used as a weapon.

   gun   bowl   spear   arrow

2. My sister likes to **wander** through the hills on her horse.

   Circle the word that means the same as "wander."

   fall   sneak   sit   roam

3. The **flexible** gymnast did a backbend with ease.

   Circle the word that means the opposite of "flexible."

   rough   old   stiff   tiny

4. The old man uses a **stout** cane when he walks down the street.

   Circle the word that means the same as "stout."

   strong   thin   short   yellow

5. Ralph used a very strong **alloy** to make the parts for his airplane.

   Circle the best answer. An alloy is a mixture of:

   plastics   metals   chemicals   plants

6. Sam is very good at tracking **game**.

   Circle the word that does **not** name a kind of game.

   deer   rabbit   corn   squirrel

7. We bought the building **materials** we needed at the lumberyard.

   Circle the items that might have been purchased.

   clothing   nails   boards   potato chips   hammer

Name _____

Favorite Things

Nature's Gifts

Everyone has a few favorite objects. Think of one object that you treasure a great deal. What is it?

_____

What is it made of?

_____

What is the object used for?

_____

Where did it come from?

_____

Why is this object so special to you?

_____

_____

_____

Draw a picture of the object in the box below.

Answer Key

Page 7
1. all of the above
2. 10
3. every night
4. grumpy
5. your lungs stop working

Page 8
A. 3
   4
   1
   2
   5
B. Drawings will vary.

Page 9
Students' records and answers will vary.

Page 12
1. He smelled the toast in the toaster.
2. Students' responses may vary, but might include: pushed the tape recorder button, turned on the light, picked up a pencil, pushed the drinking fountain button, kicked a soccer ball.
3. Students' responses may vary, but might include: draw pictures, create a time line, write in a diary or journal.
4. Students' responses will vary.

Page 13
A. 1. sense
   2. environment
   3. organism
   4. interact
   5. survive
B. Illustrations will vary.

Page 14

sour	sweet	salty
vinegar	cake	pretzel
lemon	cookie	french fries
pickle	sugar	salt

Page 17
1. Sound is created when an object vibrates.
2. The frequency of a sound is the number of sound waves created in one second.
3. a low pitch
4. Sounds are created on a guitar by plucking the strings, causing them to vibrate.

5. The tuning keys make the strings tighter or looser, which makes the sound they produced higher or lower.
6. A melody is a series of single notes. A chord is more than one note played at the same time.
7. It vibrates to make the guitar's sound louder.

Page 18
A. 1. fret
   2. vibrate
   3. interpret
   4. chord
   5. sound wave
   6. melody
   7. frequency
B. 1. Definitions will vary, but are likely to be similar to: a musical instrument with strings that you pluck or strum.
   2. noun
   3. yes or no, depending on the dictionary used

Page 19
1. bridge
2. sound hole
3. finger board
4. tuning keys
5. Frets

Page 22
1. a. **temperature:** causes rocks to expand and contract, which can cause pieces to break off
   b. **water:** freezing in cracks can break off pieces; flowing over rocks can wear away layers
   c. **animals:** burrowing animals can cause rocks to crack and break
2. rock particles, water, air, organic material
3. The smaller the soil particles, the more water the soil can hold.
4. Clay soil can hold the most water.
5. Earthworms make soil healthy by creating spaces in the soil. Then the soil can hold more water, and air can better move through the soil.
6. dark-colored or red

Page 23
A. 1. particles
   2. clay
   3. contract
   4. weathering
   5. fertile
   6. sand
   7. temperature
   8. expand
   9. burrow
B. Students' sentences will vary.

Page 24
1. T
2. F
3. T
4. T
5. F
6. T
7. F
8. F
9. T
10. T
11. F

Page 27
1. all of the above
2. weight
3. burning wood
4. length
5. gas

Page 28
A. Answers will vary, but should make sense.
   Sample answers: orange, playful, hot, cold
B. Drawings will vary, but should give a sense of each specified characteristic.

Page 29
Answers will vary. One way to check would be to have partners or groups compare responses.

Page 32
1. it teaches important skills
2. hunting
3. wild pigs
4. cheetahs
5. keep in shape
6. wrestling

Page 33

A. 5
4
8
2
10
9
7
3
1
6

B. Students' pictures and labels will vary.

Page 34

1. lion 3. bobcat
2. cheetah 4. tiger

Page 37

1. A habitat is the place where an animal lives.
2. Food, water, shelter, and a place to raise young are the four important elements of habitat.
3. Hummingbirds eat sweet nectar.
4. If birdbaths and feeders are placed on the ground, it will be hard for the birds to get away from predators such as cats.
5. Fungus can grow in the feeders and harm the birds.
6. Thick foliage makes a shrub or tree a good hiding place for birds.
7. Answers mentioned in the text include: cedar, holly, pine, crab apple, spruce, elderberry.

Page 38

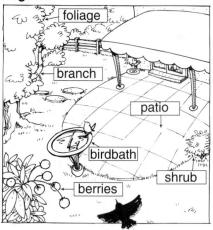

Page 39

Birds listed or drawn will vary.

Page 42

1. atoms
2. an element
3. heavy
4. move around freely
5. solid

Page 43

A. 1. molecule 5. particle
2. element 6. evaporate
3. matter 7. atoms
4. compound

B. Pictures should show that students understand the meanings of the words.

Page 44

Observations will vary, but should indicate that the microwaved ice cube shows the most rapid melting. Conclusion should state that heat speeds up the process of melting as heat makes the molecules in the water move around more freely. The greater the amount of heat applied, the faster water will change from a solid to a liquid.

Page 47

1. store
2. on the river bottom
3. the Civil War
4. so that trains could get across
5. strong

Page 48

1. cargo 5. transportation
2. foremost 6. vessel
3. steel 7. engineer
4. arches 8. successful

Page 49

Answers will vary.

Page 52

1. Limestone is formed over thousands of years by the buildup of shells and skeletons of tiny animals that lived in the seas.
2. Limestone is a sedimentary rock.
3. Limestone dissolves in water. When groundwater moves through limestone formations, the limestone dissolves over time, creating empty areas, or caves.
4. Drip vinegar on the rock. If it fizzes, it is limestone.

5. Limestone is carved into art. Limestone is cut into large blocks to build buildings. Crushed limestone is used for roads and sidewalks.
6. Millions of years ago, this land must have been at the bottom of a sea. This can be known because limestone is formed from the remains of tiny sea animals.

Page 53

1. igneous f
2. metamorphic h
3. coral b
4. quarries e
5. skeletons c
6. sedimentary i
7. dissolves d
8. magnifying a
9. limestone j
8. ancient g

Page 54

1. sedimentary rocks 4. Chapter 5
2. page 15 5. Glossary
3. Chapter 1 6. Chapter 4

Page 57

1. molten iron
2. magnet
3. a south pole and a north pole
4. Chinese navigators
5. they help us find our way
6. the Earth has a weak magnetic field

Page 58

A.
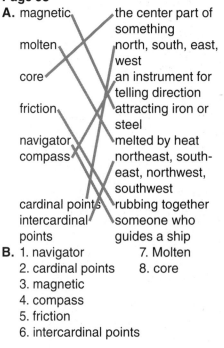

magnetic — the center part of something
molten — north, south, east, west
core — an instrument for telling direction
friction — attracting iron or steel
navigator — melted by heat
compass — northeast, southeast, northwest, southwest
cardinal points — rubbing together
intercardinal points — someone who guides a ship

B. 1. navigator 7. Molten
2. cardinal points 8. core
3. magnetic
4. compass
5. friction
6. intercardinal points

©2002 by Evan-Moor Corp.
Read and Understand, Science • Grades 3–4 • EMC 3304

Page 59
1. east
2. north
3. big rock
4. southwest
5. south
6. west

Page 62
1. Forest fires kill trees.
   Forest fires hurt or kill animals.
   Forest fires sometimes destroy homes.
2. Any three of the following:
   Forest fires clear away the litter on the forest floor.
   Forest fires provide nutrients to new and surviving trees.
   Forest fires create good habitat for birds.
   Forest fires help seeds of some plants to sprout.
   Forest fires help to get rid of non-native plants.
3. Don't play with matches.
   Be sure to put out campfires.
4. Most natural forest fires are started by lightning strikes.
5. The hot, dry weather makes the forest burn more easily. Other reasonable answers might include that there are more thunderstorms in the summer; and that more people are likely to be in the forests in the summer, making campfires and using matches.
6. Answers will vary.

Page 63
1. forest
2. litter
3. an ecologist
4. nutrients, community
5. habitat
6. snags
7. prevent, occur
8. species

Page 64
1
4
3
5
2

Page 67
1. the remains of ancient plants and animals
2. petroleum
3. all of the above
4. we are using it all up
5. all of the above

Page 68
A. 1. T   5. T   9. F
   2. T   6. T   10. F
   3. F   7. F
   4. T   8. T
B. Sources of energy include the sun, wind, petroleum, corn, and other plants.
C. Answers will vary. Accept any reasonable responses.

Page 69
A. mummies   baskets
   pitch   pools
   canoes   houses
   wagons   streets
   grease   adobe
B. Sentences will vary.

Page 72
1. Sally studied biology because she wanted to learn about the natural world. She wanted to help protect the environment.
2. Machines could not easily spin cotton with short fibers.
3. Sheets, towels, and clothing can all be made from cotton.
4. She wants to know which plants have the longest fibers and the best colors so that she can save their seeds for replanting.
5. Colored cotton does not need pesticides. It does not need dyes.

Page 73
A. 6
   10
   4
   1
   8
   5
   9
   2
   7
   3
B. Drawings and labels will vary.

Page 74
Students' stories will vary.

Page 77
1. Insulation keeps heat from moving from one place to another.
2. Ricky wants to keep the heat his body generates near his skin.

3. Each layer traps air and helps to retain the heat from Ricky's body.
4. Fat, fur, and feathers are forms of insulation found on animals' bodies.
5. Answers will vary, but may include furnace, fireplace, heaters, attic insulation, wall insulation, and double-pane windows. The last three answers would be marked with an **X**.

Page 78
1. lodge
2. chickadee
3. bound
4. insulation
5. chairlift
6. furnace
7. flitting
8. summit
9. fibers
10. plumes
11. powder
12. slope

Page 79
A potholder is used to hold hot dishes. It prevents the heat from the dish from moving to your hand and burning you.

A drink holder fits around a soft drink can. It is made of rubbery material that keeps the heat from the air from warming the drink in the can.

Earmuffs are used to keep your ears warm. They keep the heat from your ears from moving into the cold air.

Page 82
1. Monkeys and apes use branches to get termites out of the ground.
   They use rocks to crack nuts.
   They use wide leaves for shelter from rain.
2. The monkeys were able to figure out which tools were best. They chose the right tools. They did not pay attention to things that did not matter, like color.
3. a. Marc did the experiment with eggplants because he wanted to know if the monkeys understood numbers.
   b. The monkeys looked for a longer time when they did not see the number of eggplants they expected.
   c. They seemed to be able to tell when the number was wrong. They seem to have some understanding about numbers.

Page 83
A. 1. broad
   2. laboratory
   3. screen
   4. stage
   5. experiment
   6. humans
   7. termites
   8. expected
B. 1. humans
   2. termites
   3. laboratory
   4. broad
   5. expected

Page 84
Answers will vary, but should include some of the following:
Tamarins
   live in Colombia
   live in trees
   have white furry heads
   are 8 inches (20 cm) tall
Both
   eat fruit and insects
   live in groups
   are monkeys
Rhesus Monkeys
   live in India
   live on the ground
   are about 2 feet (0.6 m) tall
   are brown all over
   eat roots and seeds

Page 87
1. 11 years
2. Mercury
3. Saturn
4. Venus
5. 84 Earth years
6. Neptune

Page 88
A. 1. satellite
   2. atmosphere
   3. cycle
   4. ripples
   5. climate
   6. flare
   7. spews
   8. Exploration
   9. biome
B. Students' sentences will vary.

Page 89
Students' reports will vary.

Page 92
1. People made up stories because they did not understand the real causes of these events. Sometimes the events were frightening and the people wanted to have some explanation for why they happened.
2. During a solar eclipse, the Moon passes between the Sun and the Earth. The shadow of the Moon blocks out the Sun's light. The Moon makes a big shadow on the surface of the Earth.
3. We know that people studied the sky long ago because we have found clay tablets with records of eclipses. There are paintings in Egyptian tombs that show the planets and stars. The Greek astronomer Ptolemy understood eclipses and wrote about them.
4. A partial solar eclipse is when the Moon blocks only part of the Sun.
5. People probably have different reasons. Scientists want to learn more about eclipses. Some people are just curious. Some people find eclipses exciting.

Page 93
A. amazement—wonder
   peaceful—calm
   dim—dark
   grave—tomb
   see—view
   total—complete
   frightened—terrified
   old—ancient
   argue—disagree
B. China, Babylonia, Egypt, Greece

Page 94
Students' stories will vary.

Page 97
1. frozen rain
2. all of the above
3. thunderstorms
4. a soccer ball
5. updrafts

Page 98
1. damage
2. dent
3. shatter
4. crops
5. pelted
6. slippery
7. destroyed
8. injure
9. flattened
10. layers

Page 99

3/4 in.   4.5 cm

2 1/2 in.   1 cm

Page 102
1. the Sun
2. streaks of colorful light
3. chemicals in their bodies
4. starlight
5. they are so far away

Page 103
abdomen, firefly, zigzag, northern hemisphere, south pole, squid, aurora borealis, lightning

Page 104
Answers will vary, but natural light sources should be marked correctly. Answers to Part B should be supported.

Page 107
1. Scientists classify animals by looking at the things they have in common.
2. Aristotle was the first scientist to classify animals.
3. Aristotle grouped animals by way of living, actions, habits, and body parts.
4. Linnaeus made up a new classification system because new animals were being discovered that did not fit into Aristotle's system.
5. Scientists look at where the animals live and what they eat, as well as what the animals look like.
6. Both chipmunks and squirrels are rodents that have bushy tails.
7. Squirrels climb trees, but chipmunks don't.

Page 108
Across
   4. traits
   5. taxonomy
   7. species
   9. identify
   10. system
   11. rodents
   12. classification
Down
   1. Aristotle
   2. biologist
   3. mammals
   6. Linnaeus
   8. squirrel

Page 109
Students' classification systems will vary. Here is a sample response:

Page 112
1. wolf
2. all of the above
3. tree bark
4. there is plenty of food
5. alleys and parks
6. cities are taking up more land

Page 113
A. wilderness — an animal that hunts other animals
   adapt — a place where an animal lives
   predator — a place where no people live
   habitat — an animal that is hunted by other animals
   prey — to change because you are in a new situation
B. 1. prey   4. habitat
   2. adapt   5. wilderness
   3. predator
C. Drawings may vary.

Page 114
1. 5   4. insects
2. 20   5. Answers will vary, but
3. 5   should be justified by the graph.

Page 117
1. food
2. there isn't enough air or moisture
3. pollution
4. sorting
5. all of the above
6. it is better for the environment

Page 118
1. decompose   3. recyclables
2. contractor   4. pollution

5. environment   7. waste stream
6. recycling   8. landfill
Answer to code: Recycling reduces pollution.

Page 119
A. 4
   7
   1
   5
   2
   6
   3
B. Students' drawings will vary.

Page 122
1. is made of several large pieces
2. volcanic mountains
3. all of the above
4. large chunks of crust break and move
5. sharper and more rugged
6. Black Hills of South Dakota

Page 123
A. 1. crevice   5. magma   9. faults
   2. crust   6. cone   10. lava
   3. vent   7. dome
   4. plates   8. glacier
B. Erosion is the gradual wearing away of something by water or wind.
C. Explanations will vary, but should explain that water and moving ice carry small pieces of the mountain downhill as they move. Wind can blow away small bits of rock and soil.

Page 124
A. volcanic   folded
   fault-block   dome

Page 127
1. X-ray
2. more than 100 years old
3. bones
4. the brain
5. painless
6. help them diagnose problems

Page 128
Across	Down
1. operation	2. traditional
4. diagnose	3. patients
7. radiation	5. organs
8. dense	6. fracture
9. scanner	

Page 129
Students' answers will vary, but should answer all questions posed in the prompt.

Page 132
1. Arizona
2. a dry desert
3. warm and wet
4. mineral-rich water soaked into the trees
5. leave them where they are and tell a park ranger

Page 133
1. yes   4. yes
2. no   5. no
3. yes
Students' sentences for each item will vary.

Page 134
Fossils are of an insect, a shell, a fish, and a fern frond (or leaf). Sentences will vary.

Page 137
1. **stone:** spear point
   bark: basket
   animal skins: clothing, blankets
   wood: spears, digging sticks
2. Possible answers include:

plants	metals	plastics
oils	machinery	food containers
paint	automobiles	trash cans
soap	airplanes	football helmets
medicine	coins	parts for space
cloth	jewelry	shuttles

3. Plastics are made from chemicals found in natural materials such as petroleum and coal.
4. Answers will vary.

Page 138
1. bowl
2. roam
3. stiff
4. strong
5. metals
6. corn
7. nails, boards, hammer

Page 139
Students' responses will vary.